Student Solutions Guide to accompany
CHARLES P. MᶜKEAGUE

TRIGONOMETRY

SECOND EDITION

JUDITH M. BARCLAY

CUESTA COLLEGE

(HBJ)

Harcourt Brace Jovanovich, Publishers
and its subsidiary, Academic Press

San Diego   New York   Chicago   Austin   Washington, D.C.
London   Sydney   Tokyo   Toronto

ISBN: 0-15-592364-1
Printed in the United States of America

PREFACE TO THE STUDENT

This manual contains detailed explanations for solving every other odd problem from <u>Trignometry</u>, Second Edition, by Charles P. McKeague. There is no substitute for working through a problem on your own, but I hope that this manual will end some of the frustration involved in problem solving.

Try all the problems on your own before consulting this manual. It will also be helpful to study one problem and then use the same technique to solve a similar one whose solution is not included. The steps taken in solving a problem are the important part, not the answer.

You wouldn't think about becoming a good tennis player without a great deal of practice. The same is true in mathematics. Happy problem solving and good luck.

Judy Barclay

CONTENTS

THE SIX TRIGONOMETRIC FUNCTIONS

Problem Set 1.1

1. 10° is an acute angle.
    The complement of 10° is 80° because 10° + 80° = 90°.
    The supplement of 10° is 170° because 10° + 170° = 180°.

5. 120° is an obtuse angle.
    The complement of 120° is –30° because 120° + (–30°) = 90°.
    The supplement of 120° is 60° because 120° + 60° = 180°.

9. $\alpha = 180° - (\angle A + \angle D)$ The sum of the angles of a triangle is 180°
    $= 180° - (30° + 90°)$ Substitute given values
    $= 180° - 120°$ Simplify
    $= 60°$

13. $\angle A = 180° - (\alpha + \beta + \angle B)$ The sum of the angles of a triangle is 180°
    $= 180° - (100° + 30°)$ Substitute given values
    $= 180° - 130°$ Simplify
    $= 50°$

17. $\alpha + \beta = 90°$ α and β are complementary
    $\beta = 90° - \alpha$ Subtract α from both sides
    $\beta = 90° - 25°$ Substitute given value
    $\beta = 65°$ Simplify

21. One complete revolution equals 360°.

    In 4 hours, the hour hand revolves $\frac{4}{12}$ or $\frac{1}{3}$ of a revolution.

    $\frac{1}{3}$ of 360° = 120°.

25. $c^2 = a^2 + b^2$ Pythagorean theorem
    $c^2 = (4)^2 + (3)^2$ Substitute given values
    $= 16 + 9$ Simplify
    $= 25$

    Therefore, c = ±5. Our only solution is c = 5, because we cannot use c = –5.

29. $a^2 + b^2 = c^2$ Pythagorean theorem
    $a^2 = c^2 - b^2$ Subtract b^2 from both sides
    $a^2 = (13)^2 - (12)^2$ Substitute given values
    $= 169 - 144$ Simplify
    $= 25$

    Therefore, a = ±5. Our only solution is a = 5, because we cannot use a = –5.

33. $x^2 = (2)^2 + (2\sqrt{3})^2$ Pythagorean theorem
       $= 4 + 12$ Simplify
       $= 16$

    Therefore, $x = \pm 4$. Our only solution is $x = 4$, because we cannot use $x = -4$.

37. $(BD)^2 = (CD)^2 + (BC)^2$ Pythagorean theorem
      $5^2 = (CD)^2 + (4)^2$ Substitute given values
     $25 = (CD)^2 + 16$ Simplify
       $9 = (CD)^2$ Subtract 16 from both sides
    $CD = 3$ or ~~$CD = -3$~~   Take square root of both sides

    Therefore, $AC = 2 + 3 = 5$ $AC = AD + DC$

    $(AB)^2 = (AC)^2 + (BC)^2$ Pythagorean theorem
         $= 5^2 + 4^2$ Substitute given values
         $= 25 + 16$ Simplify
    $(AB)^2 = 41$
      $AB = \sqrt{41}$ or ~~$AB = -\sqrt{41}$~~   Take square root of both sides

41. This is an isosceles triangle. Therefore, the altitude must bisect the base.

    $x^2 = (18)^2 + (13.5)^2$ Pythagorean theorem
      $= 324 + 182.25$ Simplify
    $x^2 = 506.25$
    $x = 22.5$ or ~~$x = -22.5$~~   Take square root of both sides
    $x = 22.5$ feet

45. The longest side is 8 which is twice the shortest side.
    Therefore, the shortest side is 4.
    The side opposite the 60° angle is $4\sqrt{3}$.

49. The shortest side is 20'.
    The longest side is twice the shortest side.
    Therefore, $x = 2(20)$
            $x = 40'$

53. hypotenuse $= \dfrac{4}{5} \cdot \sqrt{2}$ Hypotenuse is $t \cdot \sqrt{2}$

            $= \dfrac{4\sqrt{2}}{5}$ Simplify

57. hypotenuse $= t\sqrt{2}$ t is the shorter side
           $4 = t\sqrt{2}$ Substitute given value

       $\dfrac{4}{\sqrt{2}} = t$ Divide both sides by $\sqrt{2}$

         $t = 2\sqrt{2}$ Rationalize denominator by multiplying numerator and denominator by $\sqrt{2}$

Problem Set 1.2

13.   If we let x = 0, the equation   3x + 2y = 6

$$becomes\ 3(0) + 2y = 6$$
$$2y = 6$$
$$y = 3$$

This gives us (0, 3) as one solution to 3x + 2y = 6.

If we let y = 0, the equation   3x + 2y = 6

$$becomes\ 3x + 2(0) = 6$$
$$3x = 6$$
$$x = 2$$

This gives us (2, 0) as a second solution to 3x + 2y = 6.

Graphing the points (0, 3) and (2, 0) and then drawing a line through them, we have the graph of 3x + 2y = 6.

17.   If we let x = 0, the equation $y = \frac{1}{2}x$

$$becomes\ y = \frac{1}{2}(0)$$

$$y = 0$$

This gives us (0, 0) as one solution to $y = \frac{1}{2}x$.

If we let x = 4, the equation $y = \frac{1}{2}x$

$$becomes\ y = \frac{1}{2}(4)$$

$$y = 2$$

This gives us (4, 2) as a second solution to $y = \frac{1}{2}x$.

Graphing the points (0, 0) and (4, 2), and then drawing a line through them, we have the graph of $y = \frac{1}{2}x$.

21.   If we let x = 0, the equation $\quad x^2 + y^2 = 1$

$$becomes\ (0)^2 + y^2 = 1$$
$$y^2 = 1$$
$$y = -1 \text{ or } y = 1$$

This gives us (0, −1) and (0, 1) as two solutions to $x^2 + y^2 = 1$.

If we let y = 0, the equation $\quad x^2 + y^2 = 1$

$$becomes\ x^2 + (0)^2 = 1$$
$$x^2 = 1$$
$$x = -1 \text{ or } x = 1$$

This gives us (−1, 0) and (1, 0) as two other solutions to $x^2 + y^2 = 1$.

Graphing the points (0, −1), (0, 1), (−1, 0), and (1, 0), and then connecting them with a smooth curve (a circle), we have the graph of $x^2 + y^2 = 1$.

25. $\begin{aligned} r &= \sqrt{(x_2 - x_1)^2 + (y_2 - y_1)^2} \qquad \text{Distance formula} \\ &= \sqrt{(0 - 5)^2 + (12 - 0)^2} \qquad \text{Substitute given values} \\ &= \sqrt{(-5)^2 + (12)^2} \qquad \text{Simplify} \\ &= \sqrt{25 + 144} \\ &= \sqrt{169} \\ &= 13 \end{aligned}$

29. $\begin{aligned} r &= \sqrt{(x_2 - x_1)^2 + (y_2 - y_1)^2} \qquad \text{Distance formula} \\ &= \sqrt{[-1 - (-10)]^2 + (-2 - 5)^2} \qquad \text{Substitute given values} \\ &= \sqrt{(9)^2 + (-7)^2} \qquad \text{Simplify} \\ &= \sqrt{81 + 49} \\ &= \sqrt{130} \end{aligned}$

33. $\begin{aligned} r &= \sqrt{(x_2 - x_1)^2 + (y_2 - y_1)^2} \qquad \text{Distance formula} \\ \sqrt{13} &= \sqrt{(x - 1)^2 + (2 - 5)^2} \qquad \text{Substitute given values} \\ \sqrt{13} &= \sqrt{(x - 1)^2 + (-3)^2} \qquad \text{Simplify} \\ \sqrt{13} &= \sqrt{(x - 1)^2 + 9} \\ 13 &= (x - 1)^2 + 9 \qquad \text{Square both sides} \\ (x - 1)^2 &= 4 \qquad \text{Subtract 9 from both sides} \\ x - 1 &= 2 \text{ or } x - 1 = -2 \qquad \text{Take square root of both sides} \\ x &= 3 \qquad\quad x = -1 \end{aligned}$

Check both solutions:

$\begin{aligned} \sqrt{13} &= \sqrt{(3 - 1)^2 + 9} \\ &= \sqrt{(2)^2 + 9} \\ &= \sqrt{4 + 9} \\ \sqrt{13} &\overset{\checkmark}{=} \sqrt{13} \end{aligned}$ $\begin{aligned} \sqrt{13} &= \sqrt{(-1 - 1)^2 + 9} \\ &= \sqrt{(-2)^2 + 9} \\ &= \sqrt{4 + 9} \\ \sqrt{13} &\overset{\checkmark}{=} \sqrt{13} \end{aligned}$

37. The baseball diamond is a square with sides of 60 feet.
The coordinates of home plate would be (0, 0).
The coordinates of first base would be (60, 0).
(60 to the right, 0 up)
The coordinates of second base would be (60, 60).
(60 to the right, 60 up)
The coordinates of third base would be (0, 60).
(0 right, 60 up)

41. Quadrants I and II lie above the x-axis. Therefore, all points in these two quadrants have positive y-coordinates.

45. a. If we draw 135° in standard position, we see that
         the terminal side is along the line y = -x.
         Since the terminal side lies in the second
         quadrant, x is negative and y is positive.
         Some points on the terminal side are:

         $(-1, 1)$, $(-\frac{1}{2}, \frac{1}{2})$, $(-3, 3)$, $(-\sqrt{2}, \sqrt{2})$, and $(-5, 5)$.

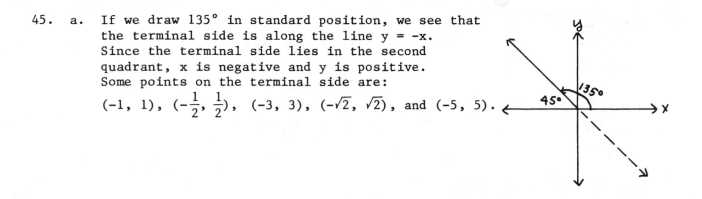

     b. To find the distance from (0, 0) to (-3, 3), we use the distance formula:

         $r = \sqrt{(x_2 - x_1)^2 + (y_2 - y_1)^2}$

         $ = \sqrt{(-3 - 0)^2 + (3 - 0)^2}$ Substitute given values

         $ = \sqrt{(-3)^2 + (3)^2}$ Simplify

         $ = \sqrt{9 + 9}$

         $ = \sqrt{18}$

         $ = 3\sqrt{2}$

     c. The angle between 0° and -360° that is coterminal with 135° is -225°

49. a. If we draw 90° in standard position, we see that the terminal side is along
         the positive y-axis.
         Some points on the terminal side are: $(0, 1)$, $(0, \frac{3}{2})$, $(0, \sqrt{2})$, $(0, 3)$ and
         $(0, 5)$.

     b. The distance from (0, 0) to (0, 1) is 1 unit. The distance from (0, 0) to
         (0, 3) is 3 units.

     c. The angle between 0° and -360° that is coterminal with 90° is -270°.

53. The figure shows the point (a, 1) on the terminal side.
     The side opposite the 60° angle is equal to the side
     opposite the 30° angle multiplied by $\sqrt{3}$.

     Therefore, $a = 1 \cdot \sqrt{3}$
     $ a = \sqrt{3}$

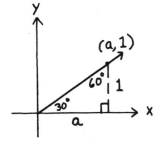

57. $a = \sqrt{(5 - 0)^2 + (0 - 0)^2}$

$= \sqrt{5^2}$

$= 5$

$b = \sqrt{(5 - 5)^2 + (12 - 0)^2}$

$= \sqrt{12^2}$

$= 12$

$c = \sqrt{(5 - 0)^2 + (12 - 0)^2}$

$= \sqrt{25 + 144}$

$= \sqrt{169}$

$= 13$

Does $a^2 + b^2 = c^2$?

$5^2 + 12^2 = 13^2$

$25 + 144 = 169$

$169 \overset{\checkmark}{=} 169$

Therefore, it is a right triangle.

Problem Set 1.3

1. $(x, y) = (3, 4)$ $\sin \theta = \dfrac{y}{r} = \dfrac{4}{5}$

$x = 3$ and $y = 4$ $\cos \theta = \dfrac{x}{r} = \dfrac{3}{5}$

$r = \sqrt{3^2 + 4^2}$ $\tan \theta = \dfrac{y}{x} = \dfrac{4}{3}$

$= \sqrt{9 + 16}$ $\cot \theta = \dfrac{x}{y} = \dfrac{3}{4}$

$= \sqrt{25}$ $\sec \theta = \dfrac{r}{x} = \dfrac{5}{3}$

$= 5$ $\csc \theta = \dfrac{r}{y} = \dfrac{5}{4}$

5. $(x, y) = (-5, 12)$ $\sin \theta = \dfrac{y}{r} = \dfrac{12}{13}$

    $x = -5$ and $y = 12$ $\cos \theta = \dfrac{x}{r} = -\dfrac{5}{13}$

    $r = \sqrt{(-5)^2 + (12)^2}$ $\tan \theta = \dfrac{y}{x} = -\dfrac{12}{5}$

       $= \sqrt{25 + 144}$ $\cot \theta = \dfrac{x}{y} = -\dfrac{5}{12}$

       $= \sqrt{169}$ $\sec \theta = \dfrac{r}{x} = -\dfrac{13}{5}$

       $= 13$ $\csc \theta = \dfrac{r}{y} = \dfrac{13}{12}$

9. $(x, y) = (a, b)$ $\sin \theta = \dfrac{y}{r} = \dfrac{b}{\sqrt{a^2 + b^2}}$

    $x = a$ and $y = b$ $\cos \theta = \dfrac{x}{r} = \dfrac{a}{\sqrt{a^2 + b^2}}$

    $r = \sqrt{a^2 + b^2}$ $\tan \theta = \dfrac{y}{x} = \dfrac{b}{a}$

       $\cot \theta = \dfrac{x}{y} = \dfrac{a}{b}$

       $\sec \theta = \dfrac{r}{x} = \dfrac{\sqrt{a^2 + b^2}}{a}$

       $\csc \theta = \dfrac{r}{y} = \dfrac{\sqrt{a^2 + b^2}}{b}$

13. $(x, y) = (\sqrt{3}, -1)$ $\sin \theta = \dfrac{y}{r} = -\dfrac{1}{2}$

    $x = \sqrt{3}$ and $y = -1$ $\cos \theta = \dfrac{x}{r} = \dfrac{\sqrt{3}}{2}$

    $r = \sqrt{(\sqrt{3})^2 + (-1)^2}$ $\tan \theta = \dfrac{y}{x} = -\dfrac{1}{\sqrt{3}}$

       $= \sqrt{3 + 1}$ $\cot \theta = \dfrac{x}{y} = -\dfrac{\sqrt{3}}{1} = -\sqrt{3}$

       $= \sqrt{4}$ $\sec \theta = \dfrac{r}{x} = \dfrac{2}{\sqrt{3}}$

       $= 2$ $\csc \theta = \dfrac{r}{y} = -\dfrac{2}{1} = -2$

17.　(x, y) = (60, 80)　　　　$\sin \theta = \dfrac{y}{r} = \dfrac{80}{100} = \dfrac{4}{5}$

　　　x = 60 and y = 80　　　$\cos \theta = \dfrac{x}{r} = \dfrac{60}{100} = \dfrac{3}{5}$

　　　$r = \sqrt{(60)^2 + (80)^2}$　　$\tan \theta = \dfrac{y}{x} = \dfrac{80}{60} = \dfrac{4}{3}$

　　　　$= \sqrt{3600 + 6400}$　　$\cot \theta = \dfrac{x}{y} = \dfrac{60}{80} = \dfrac{3}{4}$

　　　　$= \sqrt{10,000}$　　　$\sec \theta = \dfrac{r}{x} = \dfrac{100}{60} = \dfrac{5}{3}$

　　　　$= 100$　　　　　$\csc \theta = \dfrac{r}{y} = \dfrac{100}{80} = \dfrac{5}{4}$

21.　(x, y) = (9.36, 7.02)　　$\sin \theta = \dfrac{y}{r} = \dfrac{7.02}{11.7} = .6$

　　　x = 9.36 and y = 7.02

　　　$r = \sqrt{(9.36)^2 + (7.02)^2}$　　$\cos \theta = \dfrac{x}{r} = \dfrac{9.36}{11.7} = .8$

　　　　$= \sqrt{87.6096 + 49.2804}$

　　　　$= \sqrt{136.89}$

　　　　$= 11.7$

25.　A point on the terminal side of an angle of 90° is (0, 1).

　　　(x, y) = (0, 1)　　$\sin \theta = \dfrac{y}{r} = \dfrac{1}{1} = 1$

　　　x = 0 and y = 1　　$\cos \theta = \dfrac{x}{r} = \dfrac{0}{1} = 0$

　　　$r = \sqrt{0^2 + 1^2}$　　　$\tan \theta = \dfrac{y}{x} = \dfrac{1}{0}$ (undefined)

　　　　$= \sqrt{1}$

　　　　$= 1$

29.　A point on the terminal side of an angle of 0° is (1, 0).

　　　(x, y) = (1, 0)　　$\sin \theta = \dfrac{y}{r} = \dfrac{0}{1} = 0$

　　　x = 1 and y = 0　　$\cos \theta = \dfrac{x}{r} = \dfrac{1}{1} = 1$

　　　$r = \sqrt{1^2 + 0^2}$　　　$\tan \theta = \dfrac{y}{x} = \dfrac{0}{1} = 0$

　　　　$= \sqrt{1}$

　　　　$= 1$

33. $\sin \theta = \dfrac{y}{r}$ (r is always positive)

If $\sin \theta$ is negative, then y must be negative.
y is negative in quadrants III and IV.
Therefore, $\sin \theta$ is negative in Q III and Q IV.

37. $\sin \theta = \dfrac{y}{r}$ and $\tan \theta = \dfrac{y}{x}$

If $\tan \theta$ is positive, the signs of x and y must be the same. That is, x and y are positive or x and y are negative. Therefore, $\tan \theta$ is positive in Q I and Q III.
In problem 33, we found that the $\sin \theta$ is negative in Q III and Q IV.
Therefore, $\tan \theta$ is positive and $\sin \theta$ is negative in Q III.

41. $\sin \theta = \dfrac{y}{r} = \dfrac{12}{13}$ and θ terminates in Q I.

y = 12 and r = 13

$$x^2 + y^2 = r^2$$
$$x^2 + (12)^2 = (13)^2$$
$$x^2 + 144 = 169$$
$$x^2 = 25$$
$$x = \pm 5$$

Since θ terminates in Q I, x must equal 5.

$\cos \theta = \dfrac{x}{r} = \dfrac{5}{13}$

$\tan \theta = \dfrac{y}{x} = \dfrac{12}{5}$

$\cot \theta = \dfrac{x}{y} = \dfrac{5}{12}$

$\sec \theta = \dfrac{r}{x} = \dfrac{13}{5}$

$\csc \theta = \dfrac{r}{y} = \dfrac{13}{12}$

45. sin θ is negative in Q III and Q IV.
     sec θ is positive in Q I and Q IV. (See Table in text)
     Therefore, θ is in Q IV and y is negative.

$$\sec \theta = \frac{r}{x} = \frac{13}{5}$$

r = 13 and x = 5

$$x^2 + y^2 = r^2$$
$$(5)^2 + y^2 = (13)^2$$
$$25 + y^2 = 169$$
$$y^2 = 144$$
$$y = \pm 12$$

Since y is negative, y = -12.

$$\sin \theta = \frac{y}{r} = -\frac{12}{13}$$

$$\cos \theta = \frac{x}{r} = \frac{5}{13}$$

$$\tan \theta = \frac{y}{x} = -\frac{12}{5}$$

$$\cot \theta = \frac{x}{y} = -\frac{5}{12}$$

$$\csc \theta = \frac{r}{y} = -\frac{13}{12}$$

49. θ terminates in Q IV. Therefore, x is positive and y is negative.

$$\cos \theta = \frac{x}{r} = \frac{\sqrt{3}}{2}$$

Therefore, x = √3 and r = 2

$$x^2 + y^2 = r^2$$
$$(\sqrt{3})^2 + y^2 = (2)^2$$
$$3 + y^2 = 4$$
$$y^2 = 1$$
$$y = \pm 1$$

Therefore, y = -1

$$\sin \theta = \frac{y}{r} = -\frac{1}{2}$$

$$\tan \theta = \frac{y}{x} = -\frac{1}{\sqrt{3}}$$

$$\cot \theta = \frac{x}{y} = \frac{\sqrt{3}}{-1} = -\sqrt{3}$$

$$\sec \theta = \frac{r}{x} = \frac{2}{\sqrt{3}}$$

$$\csc \theta = \frac{r}{y} = \frac{2}{-1} = -2$$

53. $\tan \theta = \dfrac{y}{x} = \dfrac{a}{b}$.   Therefore, $y = a$ and $x = b$.

$r^2 = x^2 + y^2$

$r^2 = b^2 + a^2$

$r = \pm\sqrt{a^2 + b^2}$

$r = \sqrt{a^2 + b^2}$ since r is always positive

$\sin \theta = \dfrac{y}{r} = \dfrac{a}{\sqrt{a^2 + b^2}}$

$\cos \theta = \dfrac{x}{r} = \dfrac{b}{\sqrt{a^2 + b^2}}$

$\cot \theta = \dfrac{x}{y} = \dfrac{b}{a}$

$\sec \theta = \dfrac{r}{x} = \dfrac{\sqrt{a^2 + b^2}}{b}$

$\csc \theta = \dfrac{r}{y} = \dfrac{\sqrt{a^2 + b^2}}{a}$

57. $\tan \theta = \dfrac{y}{x}$ and θ is in Q III.

Therefore, x and y are negative and $\tan \theta = 1$.
Assume $y = -1$ and $x = -1$.
Then $r^2 = x^2 + y^2$
$\qquad r^2 = (-1)^2 + (-1)^2$
$\qquad r^2 = 1 + 1$
$\qquad r^2 = 2$
$\qquad\quad r = \sqrt{2}$ because r is always positive

A triangle whose sides are in this relationship must be a 45°–45°–90° right triangle.
The angle θ in standard position in Q III would be: $\theta = 180° + 45°$
$\hspace{9.5cm}\theta = 225°$

61. The terminal side of θ lies in Q II.
Let $x = -1$. Then $y = -3x$
$\hspace{3.5cm} y = -3(-1)$
$\hspace{3.5cm} y = 3$

Therefore, the point $(-1, 3)$ is on the terminal side of θ.

$r^2 = x^2 + y^2$ $\hspace{2cm}$ $\sin \theta = \dfrac{y}{r} = \dfrac{3}{\sqrt{10}}$

$r^2 = (-1)^2 + (3)^2$ $\hspace{1cm}$ $\tan \theta = \dfrac{y}{x} = \dfrac{3}{-1} = -3$

$\hspace{0.8cm} = 1 + 9$

$r^2 = 10$

$\hspace{0.3cm} r = \sqrt{10}$

65. $\sin \theta = \dfrac{y}{r} = \dfrac{-3}{5}$

Therefore, $y = -3$ and $r = 5$.

$$x^2 + y^2 = r^2$$
$$x^2 + (-3)^2 = (5)^2$$
$$x^2 + 9 = 25$$
$$x^2 = 16$$
$$x = \pm 4$$

Problem Set 1.4

1. $\dfrac{1}{7}$

5. $\dfrac{1}{-1/\sqrt{2}} = 1 \cdot \dfrac{\sqrt{2}}{-1} = -\sqrt{2}$

9. $\csc \theta = \dfrac{1}{\sin \theta}$

   $= \dfrac{1}{4/5}$

   $= \dfrac{5}{4}$

13. $\cot \theta = \dfrac{1}{\tan \theta}$

    $= \dfrac{1}{a}$, $a \neq 0$

17. $\cot \theta = \dfrac{\cos \theta}{\sin \theta}$

    $= \dfrac{-12/13}{-5/13}$

    $= \dfrac{12}{5}$

21. $\tan^3 \theta = (\tan \theta)^3$
    $= (2)^3$
    $= 8$

25. $\sec \theta = \dfrac{1}{\cos \theta}$

    $= \dfrac{1}{-5/13}$

    $= -\dfrac{13}{5}$

29. $\cos \theta = \pm\sqrt{1 - \sin^2 \theta}$

$= \pm\sqrt{1 - (\frac{1}{3})^2}$

$= \pm\sqrt{1 - \frac{1}{9}}$

$= \pm\sqrt{\frac{8}{9}}$

$= \pm \frac{2\sqrt{2}}{3}$

Also, θ terminates in Q II and $\cos \theta$ is negative in Q II.

Therefore, $\cos \theta = - \frac{2\sqrt{2}}{3}$

33. $\sin \theta = \pm\sqrt{1 - \cos^2 \theta}$

$= \pm\sqrt{1 - (\frac{\sqrt{3}}{2})^2}$

$= \pm\sqrt{1 - \frac{3}{4}}$

$= \pm\sqrt{\frac{1}{4}}$

$= \pm \frac{1}{2}$

Also, θ terminates in Q I and $\sin \theta$ is positive in Q I.

Therefore, $\sin \theta = \frac{1}{2}$

37. $\cos \theta = \pm\sqrt{1 - \sin^2 \theta}$

$= \pm\sqrt{1 - (\frac{1}{3})^2}$

$= \pm\sqrt{1 - \frac{1}{9}}$

$= \pm\sqrt{\frac{8}{9}}$

$= \pm \frac{2\sqrt{2}}{3}$

Also, θ terminates in Q I and $\cos \theta$ is positive in Q I.

Therefore, $\cos \theta = \frac{2\sqrt{2}}{3}$

$\tan \theta = \frac{\sin \theta}{\cos \theta}$

$= \frac{1/3}{2\sqrt{2}/3}$

$= \frac{1}{2\sqrt{2}}$

14

41. Sin θ is negative in Q III and Q IV, but θ is not in Q III. Therefore, θ must be in Q IV.

$$\csc \theta = \frac{1}{\sin \theta}$$

$$= \frac{1}{-1/2}$$

$$= -2$$

$$\cos \theta = \pm\sqrt{1 - \sin^2 \theta}$$

$$= \pm\sqrt{1 - (-\frac{1}{2})^2}$$

$$= \pm\sqrt{1 - \frac{1}{4}}$$

$$= \pm\sqrt{\frac{3}{4}}$$

$$= \pm\frac{\sqrt{3}}{2}$$

$\cos \theta = \frac{\sqrt{3}}{2}$ because $\cos \theta$ is positive in Q IV

$$\tan = \frac{\sin \theta}{\cos \theta}$$

$$= \frac{-1/2}{\sqrt{3}/2}$$

$$= -\frac{1}{\sqrt{3}}$$

$$\cot \theta = \frac{1}{\tan \theta}$$

$$= \frac{1}{-1/\sqrt{3}}$$

$$= -\sqrt{3}$$

$$\sec \theta = \frac{1}{\cos \theta}$$

$$= \frac{1}{\sqrt{3}/2}$$

$$= \frac{2}{\sqrt{3}}$$

All six ratios are:

$$\sin \theta = -\frac{1}{2} \qquad\qquad \cot \theta = -\sqrt{3}$$

$$\cos \theta = \frac{\sqrt{3}}{2} \qquad\qquad \sec \theta = \frac{2}{\sqrt{3}}$$

$$\tan \theta = -\frac{1}{\sqrt{3}} \qquad\qquad \csc \theta = -2$$

45. $\sin \theta = \pm\sqrt{1 - \cos^2 \theta}$

$$= \pm\sqrt{1 - \left(\frac{2}{\sqrt{13}}\right)^2}$$

$$= \pm\sqrt{1 - \frac{4}{13}}$$

$$= \pm\sqrt{\frac{9}{13}}$$

$$= \pm\frac{3}{\sqrt{13}}$$

$$= -\frac{3}{\sqrt{13}} \text{ because sin } 0 \text{ is negative in Q IV}$$

$$\tan \theta = \frac{\sin \theta}{\cos \theta} \qquad\qquad \cot \theta = \frac{1}{\tan \theta}$$

$$= \frac{-3/\sqrt{13}}{2/\sqrt{13}} \qquad\qquad = \frac{1}{-3/2}$$

$$= -\frac{3}{2} \qquad\qquad = -\frac{2}{3}$$

$$\sec \theta = \frac{1}{\cos \theta} \qquad\qquad \csc \theta = \frac{1}{\sin \theta}$$

$$= \frac{1}{2/\sqrt{13}} \qquad\qquad = \frac{1}{-3/\sqrt{13}}$$

$$= \frac{\sqrt{13}}{2} \qquad\qquad = -\frac{\sqrt{13}}{3}$$

All six ratios are:

$$\sin \theta = -\frac{3}{\sqrt{13}} \qquad\qquad \cot \theta = -\frac{2}{3}$$

$$\cos \theta = \frac{2}{\sqrt{13}} \qquad\qquad \sec \theta = \frac{\sqrt{13}}{2}$$

$$\tan \theta = -\frac{3}{2} \qquad\qquad \csc \theta = -\frac{\sqrt{13}}{3}$$

16

49. $\sin \theta = \dfrac{1}{\csc \theta}$

$\sin \theta = \dfrac{1}{a}$, $a \neq 0$

$\cos \theta = \pm\sqrt{1 - \sin^2 \theta}$

$ = \pm\sqrt{1 - (\dfrac{1}{a})^2}$

$ = \pm\sqrt{1 - \dfrac{1}{a^2}}$

$ = \pm\sqrt{\dfrac{a^2 - 1}{a^2}}$

$ = \pm\dfrac{\sqrt{a^2 - 1}}{a}$

$ = \dfrac{\sqrt{a^2 - 1}}{a}$ because θ is in Q I

$\tan \theta = \dfrac{\sin \theta}{\cos \theta}$ $\qquad\qquad$ $\cot \theta = \dfrac{1}{\tan \theta}$

$ = \dfrac{1/a}{\sqrt{a^2 - 1}/a}$ $\qquad\qquad$ $ = \dfrac{1}{1/\sqrt{a^2 - 1}}$

$ = \dfrac{1}{\sqrt{a^2 - 1}}$ $\qquad\qquad$ $ = \sqrt{a^2 - 1}$

$\sec \theta = \dfrac{1}{\cos \theta}$

$ = \dfrac{1}{\sqrt{a^2 - 1}/a}$

$ = \dfrac{a}{\sqrt{a^2 - 1}}$

All six ratios are:

$\sin \theta = \dfrac{1}{a}$ $\qquad\qquad$ $\cot \theta = \sqrt{a^2 - 1}$

$\cos \theta = \dfrac{\sqrt{a^2 - 1}}{a}$ $\qquad\qquad$ $\sec \theta = \dfrac{a}{\sqrt{a^2 - 1}}$

$\tan \theta = \dfrac{1}{\sqrt{a^2 - 1}}$ $\qquad\qquad$ $\csc \theta = a$

53. $\cos \theta = \dfrac{1}{\sec \theta}$

$= \dfrac{1}{-1.24}$

$= -0.806$

$= -0.81$ (rounded to the nearest hundredth)

$\sin \theta = \pm\sqrt{1 - \cos^2 \theta}$

$= \pm\sqrt{1 - (-.81)^2}$

$= \pm\sqrt{1 - .6561}$

$= \pm\sqrt{.3439}$

$= \pm 0.586$

$= 0.59$ because $\sin \theta$ is positive in Q II

$\tan \theta = \dfrac{\sin \theta}{\cos \theta}$ 　　　　$\cot \theta = \dfrac{1}{\tan \theta}$

$= \dfrac{0.59}{-0.81}$ 　　　　　　$= \dfrac{1}{-0.73}$

$= -0.728$ 　　　　　　　$= -1.369$

$= -0.73$ 　　　　　　　$= -1.37$

$\csc \theta = \dfrac{1}{\sin \theta}$

$= \dfrac{1}{0.59}$

$= 1.694$

$= 1.69$

All six ratios are:

$\sin \theta = 0.59$ 　　　　$\cot \theta = -1.37$
$\cos \theta = -0.81$ 　　　$\sec \theta = -1.24$
$\tan \theta = -0.73$ 　　　$\csc \theta = 1.69$

57. This line passes through $(0, 0)$ and $(1, m)$.

$\text{slope} = \dfrac{y_2 - y_1}{x_2 - x_1}$

$= \dfrac{m - 0}{1 - 0}$

$= \dfrac{m}{1}$ or m

Problem Set 1.5

1. $\cos \theta = \pm\sqrt{1 - \sin^2 \theta}$ Pythagorean identity

5. $\sec \theta = \dfrac{1}{\cos \theta}$ Reciprocal identity

9. $\csc \theta \cot \theta = \dfrac{1}{\sin \theta} \cdot \dfrac{\cos \theta}{\sin \theta}$ Reciprocal and ratio identities

    $= \dfrac{\cos \theta}{\sin^2 \theta}$ Multiplication

13. $\dfrac{\sec \theta}{\csc \theta} = \dfrac{1/\cos \theta}{1/\sin \theta}$ Reciprocal identities

    $= \dfrac{\sin \theta}{\cos \theta}$ Division of fractions

17. $\dfrac{\tan \theta}{\cot \theta} = \dfrac{\sin \theta/\cos \theta}{\cos \theta/\sin \theta}$ Ratio identities

    $= \dfrac{\sin^2 \theta}{\cos^2 \theta}$ Division of fractions

21. $\tan \theta + \sec \theta = \dfrac{\sin \theta}{\cos \theta} + \dfrac{1}{\cos \theta}$ Ratio and reciprocal identities

    $= \dfrac{\sin \theta + 1}{\cos \theta}$ Addition of fractions

25. $\sec \theta - \tan \theta \sin \theta = \dfrac{1}{\cos \theta} - \dfrac{\sin \theta}{\cos \theta} \cdot \sin \theta$ Reciprocal and ratio identities

    $= \dfrac{1}{\cos \theta} - \dfrac{\sin^2 \theta}{\cos \theta}$ Multiplication of fractions

    $= \dfrac{1 - \sin^2 \theta}{\cos \theta}$ Subtraction of fractions

    $= \dfrac{\cos^2 \theta}{\cos \theta}$ Pythagorean identity

    $= \cos \theta$ Division

29. $\dfrac{1}{\sin \theta} - \dfrac{1}{\cos \theta} = \dfrac{1}{\sin \theta} \cdot \dfrac{\cos \theta}{\cos \theta} - \dfrac{1}{\cos \theta} \cdot \dfrac{\sin \theta}{\sin \theta}$ LCD is $\sin \theta \cos \theta$

    $= \dfrac{\cos \theta}{\sin \theta \cos \theta} - \dfrac{\sin \theta}{\sin \theta \cos \theta}$ Multiplication

    $= \dfrac{\cos \theta - \sin \theta}{\sin \theta \cos \theta}$ Subtraction of fractions

33. $\dfrac{1}{\sin \theta} - \sin \theta = \dfrac{1}{\sin \theta} - \sin \theta \cdot \dfrac{\sin \theta}{\sin \theta}$ LCD is $\sin \theta$

$\qquad\qquad = \dfrac{1}{\sin \theta} - \dfrac{\sin^2 \theta}{\sin \theta}$ Multiplication

$\qquad\qquad = \dfrac{1 - \sin^2 \theta}{\sin \theta}$ Subtraction of fractions

$\qquad\qquad = \dfrac{\cos^2 \theta}{\sin \theta}$ Pythagorean identity

37. $(2 \cos \theta + 3)(4 \cos \theta - 5) = 8 \cos^2 \theta - 10 \cos \theta + 12 \cos \theta - 15$
$\qquad\qquad\qquad\qquad\qquad\quad = 8 \cos^2 \theta + 2 \cos \theta - 15$

41. $(1 - \tan \theta)(1 + \tan \theta) = 1 + \tan \theta - \tan \theta - \tan^2 \theta$
$\qquad\qquad\qquad\qquad\quad = 1 - \tan^2 \theta$

45. $(\sin \theta - 4)^2 = (\sin \theta - 4)(\sin \theta - 4)$
$\qquad\qquad\quad = \sin^2 \theta - 4 \sin \theta - 4 \sin \theta + 16$
$\qquad\qquad\quad = \sin^2 \theta - 8 \sin \theta + 16$

49. $\sin \theta \sec \theta \cot \theta = \sin \theta \cdot \dfrac{1}{\cos \theta} \cdot \dfrac{\cos \theta}{\sin \theta}$ Reciprocal and ratio identities

$\qquad\qquad\qquad = \dfrac{\sin \theta \cos \theta}{\cos \theta \sin \theta}$ Multiplication

$\qquad\qquad\qquad = 1$ Division

53. $\dfrac{\csc \theta}{\cot \theta} = \dfrac{1/\sin \theta}{\cos \theta / \sin \theta}$ Reciprocal and ratio identities

$\qquad\quad = \dfrac{\sin \theta}{\sin \theta \cos \theta}$ Division of fractions

$\qquad\quad = \dfrac{1}{\cos \theta}$ Division of like factor

$\qquad\quad = \sec \theta$ Reciprocal identity

57. $\dfrac{\sec \theta \cot \theta}{\csc \theta} = \dfrac{\dfrac{1}{\cos \theta} \cdot \dfrac{\cos \theta}{\sin \theta}}{\dfrac{1}{\sin \theta}}$ Reciprocal and ratio identities

$\qquad\qquad = \dfrac{\cos \theta \sin \theta}{\cos \theta \sin \theta}$ Division of fractions

$\qquad\qquad = 1$ Division of like factors

61. $\tan \theta + \cot \theta = \dfrac{\sin \theta}{\cos \theta} + \dfrac{\cos \theta}{\sin \theta}$ Ratio and reciprocal identities

$\qquad\qquad\qquad = \dfrac{\sin \theta}{\cos \theta} \cdot \dfrac{\sin \theta}{\sin \theta} + \dfrac{\cos \theta}{\sin \theta} \cdot \dfrac{\cos \theta}{\cos \theta}$ LCD is $\sin\theta\cos\theta$

$\qquad\qquad\qquad = \dfrac{\sin^2 \theta}{\sin \theta \cos \theta} + \dfrac{\cos^2 \theta}{\sin \theta \cos \theta}$ Multiplication

$\qquad\qquad\qquad = \dfrac{\sin^2\theta + \cos^2\theta}{\sin \theta \cos \theta}$ Addition of fractions

$\qquad\qquad\qquad = \dfrac{1}{\sin \theta \cos \theta}$ Pythagorean identity

$\qquad\qquad\qquad = \dfrac{1}{\sin \theta} \cdot \dfrac{1}{\cos \theta}$ Multiplication of fractions

$\qquad\qquad\qquad = \csc \theta \sec \theta$ Reciprocal identities

65. $\csc \theta \tan \theta - \cos \theta = \dfrac{1}{\sin \theta} \cdot \dfrac{\sin \theta}{\cos \theta} - \cos \theta$ Reciprocal and ratio identities

$\qquad\qquad\qquad = \dfrac{\sin \theta}{\sin \theta \cos \theta} - \cos \theta$ Multiplication

$\qquad\qquad\qquad = \dfrac{1}{\cos \theta} - \cos \theta$ Division of like factors

$\qquad\qquad\qquad = \dfrac{1}{\cos \theta} - \cos \theta \cdot \dfrac{\cos \theta}{\cos \theta}$ LCD is $\cos \theta$

$\qquad\qquad\qquad = \dfrac{1}{\cos \theta} - \dfrac{\cos^2\theta}{\cos \theta}$ Multiplication

$\qquad\qquad\qquad = \dfrac{1 - \cos^2\theta}{\cos \theta}$ Subtraction of fractions

$\qquad\qquad\qquad = \dfrac{\sin^2\theta}{\cos \theta}$ Pythagorean identity

69. $(\sin \theta + 1)(\sin \theta - 1) = \sin^2\theta - 1$ Multiplication

$\qquad\qquad\qquad\qquad\quad = 1 - \cos^2\theta - 1$ Pythagorean identity

$\qquad\qquad\qquad\qquad\quad = -\cos^2\theta$

73. $(\sin \theta - \cos \theta)^2 - 1 = \sin^2 \theta - 2 \sin \theta \cos \theta + \cos^2 \theta - 1$

$\qquad\qquad\qquad\qquad\quad = -2 \sin \theta \cos \theta + (\sin^2 \theta + \cos^2 \theta) - 1$

$\qquad\qquad\qquad\qquad\quad = -2 \sin \theta \cos \theta + 1 - 1$

$\qquad\qquad\qquad\qquad\quad = -2 \sin \theta \cos \theta$

77. $\sin \theta (\sec \theta + \cot \theta) = \sin \theta \sec \theta + \sin \theta \cot \theta$ Multiplication

$\qquad\qquad\qquad = \sin \theta \cdot \dfrac{1}{\cos\theta} + \sin\theta \cdot \dfrac{\cos \theta}{\sin \theta}$ Reciprocal and ratio identities

$\qquad\qquad\qquad = \dfrac{\sin \theta}{\cos \theta} + \dfrac{\sin \theta \cos \theta}{\sin \theta}$ Multiplication

$\qquad\qquad\qquad = \tan \theta + \cos \theta$ Ratio identity and division of like factors

Chapter 1 Test

1. $x^2 + (3)^2 = (6)^2$ Pythagorean theorem
   $x^2 + 9 = 36$ Simplify
   $x^2 = 27$ Subtract 9 from both sides
   $x = \pm 3\sqrt{3}$ Take square root of both sides
   $x = 3\sqrt{3}$ because x must be positive

5. $(AB)^2 + (BC)^2 = (AC)^2$ Pythagorean theorem
   $(AB)^2 + (3)^2 = (5)^2$ Substitute given values
   $(AB)^2 + 9 = 25$ Simplify
   $(AB)^2 = 16$ Subtract 9 from both sides
   $AB = \pm 4$ Take square root of both sides
   $AB = 4$ because it must be positive

   $(DB)^2 = (DA)^2 + (AB)^2$ Pythagorean theorem
   $(DB)^2 = (6)^2 + (4)^2$ Substitute given values
   $(DB)^2 = 36 + 16$ Simplify
   $(DB)^2 = 52$
   $DB = \pm\sqrt{52}$ Take square root of both sides
   $DB = \sqrt{52}$ or   7.21   because it must be positive

9. $r = \sqrt{(x_2 - x_1)^2 + (y_2 - y_1)^2}$ Distance formula

   $r = \sqrt{(4 + 1)^2 + (-2 - 10)^2}$ Substitute given values

   $r = \sqrt{(5)^2 + (-12)^2}$ Simplify

   $r = \sqrt{25 + 144}$

   $r = \sqrt{169}$

   $r = 13$

13. A point on the terminal side of $-45°$ is $(1, -1)$ and $r = \sqrt{2}$.

    $\sin(-45°) = \dfrac{y}{r} = \dfrac{-1}{\sqrt{2}} = -\dfrac{1}{\sqrt{2}}$

    $\cos(-45°) = \dfrac{x}{r} = \dfrac{1}{\sqrt{2}}$

    $\tan(-45°) = \dfrac{y}{x} = \dfrac{-1}{1} = -1$

17. $(x, y) = (-3, -1)$
    $x = -3$ and $y = -1$

    $r^2 = x^2 + y^2$
    $r^2 = (-3)^2 + (-1)^2$
    $r^2 = 9 + 1$
    $r^2 = 10$
    $r = \pm\sqrt{10}$
    $r = \sqrt{10}$ because r is always positive

$$\sin \theta = \frac{y}{r} = \frac{-1}{\sqrt{10}} = -\frac{1}{\sqrt{10}}$$

$$\cot \theta = \frac{x}{y} = \frac{-3}{-1} = 3$$

$$\cos \theta = \frac{x}{r} = \frac{-3}{\sqrt{10}} = -\frac{3}{\sqrt{10}}$$

$$\sec \theta = \frac{r}{x} = \frac{\sqrt{10}}{-3} = -\frac{\sqrt{10}}{3}$$

$$\tan \theta = \frac{y}{x} = \frac{-1}{-3} = \frac{1}{3}$$

$$\csc \theta = \frac{r}{y} = \frac{\sqrt{10}}{-1} = -\sqrt{10}$$

21. $\csc \theta = \dfrac{1}{\sin \theta}$ Reciprocal identity

$\qquad = \dfrac{1}{-3/4}$ Substitute given values

$\qquad = -\dfrac{4}{3}$ Simplify

25. $\cos \theta = \pm\sqrt{1 - \sin^2 \theta}$ Pythagorean identity

$\qquad = \pm\sqrt{1 - \left(\dfrac{1}{a}\right)^2}$ Substitute given values

$\qquad = \pm\sqrt{1 - \dfrac{1}{a^2}}$ Simplify

$\qquad = \pm\sqrt{\dfrac{a^2 - 1}{a^2}}$

$\qquad = \pm\dfrac{\sqrt{a^2 - 1}}{a}$

$\qquad = \dfrac{\sqrt{a^2 - 1}}{a}$ because $\cos \theta$ is positive in Q I

$\csc \theta = \dfrac{1}{\sin \theta}$ Reciprocal identity

$\qquad = \dfrac{1}{1/a}$ Substitute given values

$\qquad = a$ Simplify

$\cot \theta = \dfrac{\cos \theta}{\sin \theta}$ Ratio identity

$\qquad = \dfrac{\sqrt{a^2 - 1}/a}{1/a}$ Substitute given values

$\qquad = \sqrt{a^2 - 1}$ Simplify

29. $\dfrac{\cot \theta}{\csc \theta} = \dfrac{\cos \theta/\sin \theta}{1/\sin \theta}$ Ratio and reciprocal identities

$\qquad = \dfrac{\cos \theta \sin \theta}{\sin \theta}$ Division of fractions

$\qquad = \cos \theta$ Division of like factors

RIGHT TRIANGLE TRIGONOMETRY

Problem Set 2.1

1. $a = \sqrt{c^2 - b^2}$ Pythagorean theorem

    $a = \sqrt{(5)^2 - (3)^2}$ Substitute given values

    $a = \sqrt{25 - 9}$ Simplify

    $a = \sqrt{16}$

    $a = 4$

$$\sin A = \frac{a}{c} = \frac{4}{5}$$

$$\cos A = \frac{b}{c} = \frac{3}{5}$$

$$\tan A = \frac{a}{b} = \frac{4}{3}$$

$$\cot A = \frac{b}{a} = \frac{3}{4}$$

$$\sec A = \frac{c}{b} = \frac{5}{3}$$

$$\csc A = \frac{c}{a} = \frac{5}{4}$$

5. $c = \sqrt{a^2 + b^2}$ Pythagorean theorem

    $c = \sqrt{(2)^2 + (\sqrt{5})^2}$ Substitute given values

    $c = \sqrt{4 + 5}$ Simplify

    $c = \sqrt{9}$

    $c = 3$

$$\sin A = \frac{a}{c} = \frac{2}{3}$$

$$\cos A = \frac{b}{c} = \frac{\sqrt{5}}{3}$$

$$\tan A = \frac{a}{b} = \frac{2}{\sqrt{5}}$$

$$\cot A = \frac{b}{a} = \frac{\sqrt{5}}{2}$$

$$\sec A = \frac{c}{b} = \frac{3}{\sqrt{5}}$$

$$\csc A = \frac{c}{a} = \frac{3}{2}$$

9. $c = \sqrt{a^2 + b^2}$ Pythagorean theorem

    $c = \sqrt{(1)^2 + (1)^2}$ Substitute given values

    $c = \sqrt{1 + 1}$ Simplify

    $c = \sqrt{2}$

$\sin A = \dfrac{a}{c} = \dfrac{1}{\sqrt{2}}$ $\sin B = \dfrac{b}{c} = \dfrac{1}{\sqrt{2}}$

$\cos A = \dfrac{b}{c} = \dfrac{1}{\sqrt{2}}$ $\cos B = \dfrac{a}{c} = \dfrac{1}{\sqrt{2}}$

$\tan A = \dfrac{a}{b} = \dfrac{1}{1} = 1$ $\tan B = \dfrac{b}{a} = \dfrac{1}{1} = 1$

13. $a = \sqrt{c^2 - b^2}$ Pythagorean theorem

    $a = \sqrt{(2x)^2 - (x)^2}$ Substitute given values

    $a = \sqrt{4x^2 - x^2}$ Simplify

    $a = \sqrt{3x^2}$

    $a = x\sqrt{3}$

$\sin A = \dfrac{a}{c} = \dfrac{x\sqrt{3}}{2x} = \dfrac{\sqrt{3}}{2}$ $\sin B = \dfrac{b}{c} = \dfrac{x}{2x} = \dfrac{1}{2}$

$\cos A = \dfrac{b}{c} = \dfrac{x}{2x} = \dfrac{1}{2}$ $\cos B = \dfrac{a}{c} = \dfrac{x\sqrt{3}}{2x} = \dfrac{\sqrt{3}}{2}$

$\tan A = \dfrac{a}{b} = \dfrac{x\sqrt{3}}{x} = \sqrt{3}$ $\tan B = \dfrac{b}{a} = \dfrac{x}{x\sqrt{3}} = \dfrac{1}{\sqrt{3}}$

17. $\tan 8° = \cot (90° - 8°)$

    $= \cot 82°$

21. $\sin x° = \cos (90° - x°)$

25. $4 \sin 30° = 4\left(\dfrac{1}{2}\right)$

    $= 2$

29. $\sin 30° \cos 45° = \left(\dfrac{1}{2}\right)\left(\dfrac{\sqrt{2}}{2}\right)$

    $= \dfrac{\sqrt{2}}{4}$

33. $\sin^2 45° - 2 \sin 45° \cos 45° + \cos^2 45°$

    $= \left(\dfrac{\sqrt{2}}{2}\right)^2 - 2\left(\dfrac{\sqrt{2}}{2}\right)\left(\dfrac{\sqrt{2}}{2}\right) + \left(\dfrac{\sqrt{2}}{2}\right)^2$

    $= \dfrac{2}{4} - 2\left(\dfrac{2}{4}\right) + \dfrac{2}{4}$

    $= \dfrac{1}{2} - 1 + \dfrac{1}{2}$

    $= 0$

37. $2 \sin 30° = 2\left(\frac{1}{2}\right)$
      $= 1$

41. $-3 \sin 2(30°) = -3 \sin 60°$
         $= -3\left(\frac{\sqrt{3}}{2}\right)$
         $= -\frac{3\sqrt{3}}{2}$

45. $\sec 30° = \dfrac{1}{\cos 30°}$ Reciprocal identity

     $= \dfrac{1}{\sqrt{3}/2}$ Substitute exact value from Table 2-2

     $= \dfrac{2}{\sqrt{3}}$ Division of fractions

49. $\cot 45° = \dfrac{\cos 45°}{\sin 45°}$ Ratio identity

     $= \dfrac{\sqrt{2}/2}{\sqrt{2}/2}$ Substitute values from Table 2-2

     $= 1$ Simplify

65. $r = \sqrt{(x_2 - x_1)^2 + (y_2 - y_1)^2}$ Distance formula

   $= \sqrt{(5 - 2)^2 + (1 - 5)^2}$ Substitute given values

   $= \sqrt{(3)^2 + (-4)^2}$ Simplify

   $= \sqrt{9 + 16}$

   $= \sqrt{25}$

   $= 5$

69. If $x = 0$, then $y = 2(0) - 1$
         $y = -1$
Therefore, the point $(0, -1)$ satisfies the equation.
If $x = 2$, then $y = 2(2) - 1$
         $y = 4 - 1$
         $y = 3$
Therefore, the point $(2, 3)$ satisfies the equation.

Plot the points $(0, -1)$ and $(2, 3)$ and draw the line through these points.

73. $-135° + 360° = 225°$

Problem Set 2.2

1. $37°45'$
   $+ 26°24'$
   $63°69' = 64°9'$ since $60' = 1°$

5. $61°33'$
   $+ 45°16'$
   $106°49'$

9. $\begin{array}{r} 180° \\ - 120°17' \end{array} = \begin{array}{r} 179°60' \\ - 120°17' \\ \hline 59°43' \end{array}$ Change 1° to 60'

13. $\begin{array}{r} 70°40' \\ - 30°50' \end{array} = \begin{array}{r} 69°100' \\ - 30°50' \\ \hline 39°50' \end{array}$ Change 1° to 60'

17. $16.25° = 16° + .25(60)'$
$= 16° + 15'$
$= 16°15'$

21. $19.9° = 19° + .9(60)'$
$= 19° + 54'$
$= 19°54'$

25. $62°36' = 62 + \dfrac{36}{60}$

$= 62.6°$

Calculator: 36 $\boxed{\div}$ 60 $\boxed{+}$ 62 $\boxed{=}$

29. $48°27' = 48 + \dfrac{27}{60}$

$= 48.45°$

Calculator: 27 $\boxed{\div}$ 60 $\boxed{+}$ 48 $\boxed{=}$

<u>Put</u> <u>calculator</u> <u>in</u> <u>degree</u> <u>mode</u>:

33. Calculator: 18 $\boxed{\cos}$
Answer to 4 places: 0.9511

37. $\cot 31° = \dfrac{1}{\tan 31°}$
Calculator: 31 $\boxed{\tan}$ $\boxed{1/x}$
Answer: 1.6643

41. $\csc 14.15° = \dfrac{1}{\sin 14.15°}$
Calculator: 14.15 $\boxed{\sin}$ $\boxed{1/x}$
Answer: 4.0906

45. $42°15' = 42 + \dfrac{15}{60}$

$= 42.25°$

Calculator: 15 $\boxed{\div}$ 60 $\boxed{+}$ 42 $\boxed{=}$

Calculator: 42.25 $\boxed{\tan}$
Answer: 0.9083

49. $88°18' = 88 + \dfrac{18}{60}$

$= 88.3°$

Calculator: 18 $\boxed{\div}$ 60 $\boxed{+}$ 88 $\boxed{=}$

$\cot 88.3° = \dfrac{1}{\tan 88.3°}$
Calculator: 88.3 $\boxed{\tan}$ $\boxed{1/x}$
Answer: 0.0297

53. $45°54' = 45 + \dfrac{54}{60}$

    Calculator: 54 $\boxed{\div}$ 60 $\boxed{+}$ 45 $\boxed{=}$

    $= 45.9°$

    $\sec 45.9° = \dfrac{1}{\cos 45.9°}$

    Calculator: 45.9 $\boxed{\cos}$ $\boxed{1/x}$

    Answer: 1.4370

57. Calculator: 0.6873 $\boxed{\text{inv}}$ $\boxed{\tan}$

    Answer: 34.5°

61. $\sec \theta = 1.0191$

    $\dfrac{1}{\cos \theta} = 1.0191$

    $\cos \theta = \dfrac{1}{1.0191}$

    Calculator: $\boxed{1/x}$ 1.0191 $\boxed{\text{inv}}$ $\boxed{\cos}$

    Answer: 11.1°

65. $\cot \theta = 0.6873$

    $\dfrac{1}{\tan \theta} = 0.6873$

    $\tan \theta = \dfrac{1}{0.6873}$

    Calculator: $\boxed{1/x}$ 0.6873 $\boxed{\text{inv}}$ $\boxed{\tan}$

    Answer: 55.5°

69. Calculator: 0.4112 $\boxed{\text{inv}}$ $\boxed{\cos}$

    Answer in decimal degrees is 65.719°

    Convert the decimal part to minutes:

    .719 $\boxed{\text{x}}$ 60 $\boxed{=}$

    To the nearest minute we have $\theta = 65°43'$

73. csc θ = 3.9451

$\dfrac{1}{\sin\ \theta}$ = 3.9451

sin θ = $\dfrac{1}{3.9451}$

Calculator: $\boxed{1/x}$ 3.9451 $\boxed{\text{inv}}$ $\boxed{\text{sin}}$
Answer in decimal degrees is 14.683°

Convert the decimal part to minutes:
.683 $\boxed{x}$ 60 $\boxed{=}$
To the nearest minute we have θ = 14°41'

77. sec θ = 1.0129

$\dfrac{1}{\cos\ \theta}$ = 1.0129

cos θ = $\dfrac{1}{1.0129}$

Calculator: $\boxed{1/x}$ 1.0129 $\boxed{\text{inv}}$ $\boxed{\text{cos}}$
Answer in decimal degrees is 9.154°

Convert the decimal part to minutes:
.154 $\boxed{x}$ 60 $\boxed{=}$
To the nearest minute we have 9°9'

81. To calculate sec 34.5° = $\dfrac{1}{\cos\ 34.5°}$:

34.5 $\boxed{\text{cos}}$ $\boxed{1/x}$

To calculate csc 55.5° = $\dfrac{1}{\sin\ 55.5°}$:

55.5 $\boxed{\text{sin}}$ $\boxed{1/x}$

Both answers should be: 1.2134

85. Calculator: 37 $\boxed{\text{cos}}$ $\boxed{x^2}$ $\boxed{+}$ 37 $\boxed{\text{sin}}$ $\boxed{x^2}$ $\boxed{=}$

89. Calculator: 1.234 $\boxed{\text{inv}}$ $\boxed{\text{sin}}$
You should get an error message. The sine of an angle can never be greater than 1.

97. A point on the terminal side of an angle of 90° in standard position is (0, 1), where x = 0, y = 1, and r = 1.

$$\sin 90° = \frac{y}{r} = \frac{1}{1} = 1$$

$$\cos 90° = \frac{x}{r} = \frac{0}{1} = 0$$

$$\tan 90° = \frac{y}{x} = \frac{1}{0} \text{ (undefined)}$$

101. sin θ is positive in Q I and Q II.
cos θ is negative in Q II and Q III.
Therefore, θ must lie in Q II.

Problem Set 2.3

1. $\sin 42° = \dfrac{a}{15}$ Sine relationship

     $a = 15 \sin 42°$ Multiply both sides by 15

     $a = 15(0.6691)$ Substitute value for sin 42°

     $a = 10$ ft   Answer rounded to 2 significant digits

5. $\cos 24.5° = \dfrac{a}{2.34}$ Cosine relationship

     $a = 2.34 \cos 24.5°$ Multiply both sides by 2.34

     $a = 2.34(0.9100)$ Substitute value for cos 24.5°

     $a = 2.13$ ft   Answer rounded to 3 significant digits

9. $\tan A = \dfrac{16}{26}$ Tangent relationship

     $\tan A = 0.6154$ Divide 16 by 26

     $A = 32°$ Answer rounded to the nearest degree

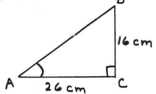

13. $\cos B = \dfrac{23.32}{45.54}$ Cosine relationship

     $\cos B = 0.5121$ Divide 23.32 by 45.54

     $B = 59.20°$ Answer rounded to nearest hundredth of a degree

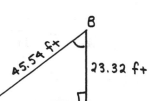

17. First, we find ∠B: ∠B = 90° - ∠A
                            ∠B = 90° - 32.6°
                            ∠B = 57.4°

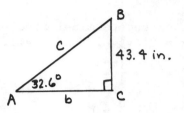

Next, we find side c:

$$\sin 32.6° = \frac{43.4}{c}$$ Sine relationship

$$c = \frac{43.4}{\sin 32.6°}$$ Multiply both sides by c and then divide by sin 32.6°

$$c = 80.6 \text{ in.}$$ Answer rounded to 3 significant digits

Last, we find side b:

$$\tan 57.4° = \frac{b}{43.4}$$ Tangent relationship

$$b = 43.4 \tan 57.4°$$ Multiply both sides by 43.4

$$b = 67.9 \text{ in.}$$ Answer rounded to 3 significant digits

21. First, we find ∠A: ∠A = 90° - 76°
                            ∠A = 14°

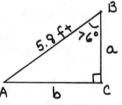

Next, we find side a:

$$\cos 76° = \frac{a}{5.8}$$ Cosine relationship

$$a = 5.8 \cos 76°$$ Multiply both sides by 5.8

$$a = 1.4 \text{ ft}$$ Answer rounded to 2 significant digits

Last, we find side b:

$$\sin 76° = \frac{b}{5.8}$$ Sine relationship

$$b = 5.8 \sin 76°$$ Multiply both sides by 5.8

$$b = 5.6 \text{ ft}$$ Answer rounded to 2 significant digits

25. First, we find ∠A: ∠A = 90° − 23.45°
                            ∠A = 66.55°

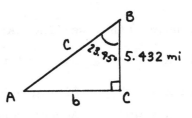

Next, we find side b:

$$\tan 23.45° = \frac{b}{5.432}$$ Tangent relationship

           b = 5.432 tan 23.45°   Multiply both sides by 5.432

           b = 2.356 mi   Answer rounded to 4 significant digits

Last, we find side c:

$$\cos 23.45° = \frac{5.432}{c}$$ Cosine relationship

$$c = \frac{5.432}{\cos 23.45°}$$ Multiply both sides by c and then divide by cos 23.45°

           c = 5.921 mi   Answer rounded to 4 significant digits

29. First, we find ∠A:

$$\sin A = \frac{2.75}{4.05}$$ Sine relationship

       sin A = 0.6790   Divide 2.75 by 4.05

           A = 42.8°   Answer rounded to the nearest tenth of a degree

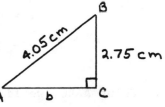

Next, we find ∠B: ∠B = 90° − 42.8°
                       ∠B = 47.2°

Last, we find side b:

$b^2 + (2.75)^2 = (4.05)^2$ Pythagorean theorem
$b^2 + 7.5625 = 16.4025$ Simplify
$b^2 = 8.84$ Subtract 7.5625 from both sides
$b = ±2.97$ Take square root of both sides
$b = 2.97$ cm   b must be positive

33. Using ∆BCD, we find BD:

$$\sin 30° = \frac{BD}{6}$$ Sine relationship

       BD = 6 sin 30°   Multiply both sides by 6

       BD = 3   Exact answer

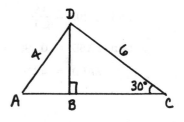

Next, using △ABD, we find ∠A:

$$\sin A = \frac{3}{4}$$ Sine relationship

$$\sin A = .75$$ Divide 3 by 4

$$A = 49°$$ Answer rounded to the nearest degree

37.

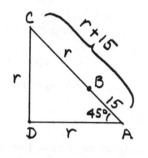

$$\sin 45° = \frac{r}{r + 15}$$ Sine relationship

$$r = (r + 15) \sin 45°$$ Multiply both sides by r + 15
$$r = r \sin 45° + 15 \sin 45°$$ Use distributive property
$$r - r \sin 45° = 15 \sin 45°$$ Subtract r sin 45° from both sides
$$r(1 - \sin 45°) = 15 \sin 45°$$ Factor out r from left side

$$r = \frac{15 \sin 45°}{1 - \sin 45°}$$ Divide both sides by 1 - sin 45°

$$r = 36$$ Answer rounded to 2 significant digits

41. Using △ABC, we find side x:

$$\sin 41° = \frac{x}{32}$$ Sine relationship

$$x = 32 \sin 41°$$ Multiply both sides by 32
$$x = 21$$ Answer rounded to 2 significant digits

Next, using △ABD, we find ∠ABD:

$$\tan ∠ABD = \frac{h}{x}$$ Tangent relationship

$$\tan ∠ABD = \frac{19}{21}$$ Substitute given values

$$\tan ∠ABD = 0.9048$$ Divide 19 by 21
$$∠ABD = 42°$$ Answer rounded to the nearest degree

45. Using ΔABC, we find side h:

$$\sin 41° = \frac{h}{28}$$ Sine relationship

$$h = 28 \sin 41°$$ Multiply both sides by 28

$$h = 18$$ Answer rounded to 2 significant digits

Next, using ΔBCD, we find side x:

$$\tan 58° = \frac{h}{x}$$ Tangent relationship

$$\tan 58° = \frac{18}{x}$$ Substitute value found for h

$$x = \frac{18}{\tan 58°}$$ Multiply both sides by x and divide by tan 58°

$$x = 11$$ Answer rounded to 2 significant digits

49. $$\cos B = \frac{1}{\sec B}$$ Reciprocal identity

$$= \frac{1}{2}$$ Substitute given value

$$\cos^2 B = \left(\frac{1}{2}\right)^2$$ Square both sides of equation

$$= \frac{1}{4}$$ Simplify

53. $$\sin A = \pm \sqrt{1 - \cos^2 A}$$ Pythagorean identity

$$= \pm \sqrt{1 - \left(\frac{2}{5}\right)^2}$$ Substitute given value

$$= \pm \sqrt{1 - \frac{4}{25}}$$ Simplify

$$= \pm \sqrt{\frac{21}{25}}$$

$$= \pm \frac{\sqrt{21}}{5}$$

However, sin A is negative in Q IV.
Therefore, $\sin A = -\dfrac{\sqrt{21}}{5}$

34

57. $\cos \theta = \dfrac{1}{\sec \theta}$ Reciprocal identity

$\quad\quad\quad = \dfrac{1}{-2} = -\dfrac{1}{2}$ Substitute given value

$\sin \theta = \pm \sqrt{1 - \cos^2 \theta}$ Pythagorean identity

$\quad\quad\quad = \pm \sqrt{1 - (-\dfrac{1}{2})^2}$ Substitute given value

$\quad\quad\quad = \pm \sqrt{1 - \dfrac{1}{4}}$ Simplify

$\quad\quad\quad = \pm \sqrt{\dfrac{3}{4}}$

$\quad\quad\quad = \pm \dfrac{\sqrt{3}}{2}$

However, $\sin \theta$ is negative in Q III

Therefore, $\sin \theta = -\dfrac{\sqrt{3}}{2}$

$\tan \theta = \dfrac{\sin \theta}{\cos \theta}$ Ratio identity

$\quad\quad\quad = \dfrac{-\sqrt{3}/2}{-1/2}$ Substitute given values

$\quad\quad\quad = \sqrt{3}$ Simplify

$\csc \theta = \dfrac{1}{\sin \theta}$ Reciprocal identity

$\quad\quad\quad = \dfrac{1}{-\sqrt{3}/2}$ Substitute given value

$\quad\quad\quad = -\dfrac{2}{\sqrt{3}}$ Simplify

$\cot \theta = \dfrac{1}{\tan \theta}$ Reciprocal identity

$\quad\quad\quad = \dfrac{1}{\sqrt{3}}$ Substitute given value

All six trigonometric ratios are:

$\sin \theta = -\dfrac{\sqrt{3}}{2}$ $\cot \theta = \dfrac{1}{\sqrt{3}}$

$\cos \theta = -\dfrac{1}{2}$ $\sec \theta = -2$

$\tan \theta = \sqrt{3}$ $\csc \theta = -\dfrac{2}{\sqrt{3}}$

Problem Set 2.4

1. To find the height, h, we can use the Pythagorean theorem:

$$h^2 + (15)^2 = (42)^2$$
$$h^2 + 225 = 1764$$
$$h^2 = 1539$$
$$h = \pm\sqrt{1539}$$
$$h = 39 \text{ cm}$$

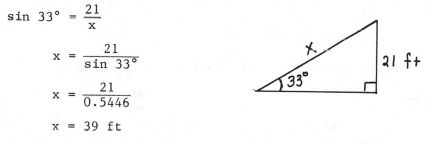

To find angle θ, we can use the cosine ratio:

$$\cos \theta = \frac{15}{42}$$

$$\cos \theta = 0.3571$$

$$\theta = 69°$$

5. To find the length of the escalator, x, we can use the sine ratio:

$$\sin 33° = \frac{21}{x}$$

$$x = \frac{21}{\sin 33°}$$

$$x = \frac{21}{0.5446}$$

$$x = 39 \text{ ft}$$

9. We use the tangent ratio to find the angle of elevation of the sun, θ:

$$\tan \theta = \frac{73.0}{51.0}$$

$$\tan \theta = 1.4314$$

$$\theta = 55.1°$$

13. To find how far the boat is from the harbor entrance, we can use the Pythagorean theorem:

$$x^2 = (25)^2 + (18)^2$$
$$x^2 = 625 + 324$$
$$x^2 = 949$$
$$x = \pm\sqrt{949}$$
$$x = 31 \text{ mi}$$

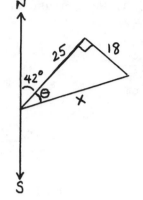

To find the bearing of the boat from the harbor entrance, first we must find angle θ using the tangent ratio:

$$\tan \theta = \frac{18}{25}$$

$$\tan \theta = 0.72$$

$$\theta = 36°$$

The bearing from the north is then $42° + 36° = 78°$ or N 78° E.

17. To find the distance traveled north, y, we use the cosine ratio:

$$\cos 37°10' = \frac{y}{79.5}$$

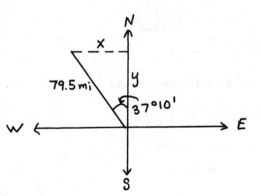

$$y = 79.5 \cos 37°10'$$
$$y = 79.5 \cos 37.1\overline{6}°$$
$$y = 79.5 (0.7969)$$
$$y = 63.4 \text{ mi north}$$

To find the distance traveled west, x, we use the sine ratio:

$$\sin 37°10' = \frac{x}{79.5}$$

$$x = 79.5 \sin 37°10'$$
$$x = 79.5 \sin 37.1\overline{6}°$$
$$x = 79.5 (0.6041)$$
$$x = 48.0 \text{ mi west}$$

21. We find h using △ACD and also using △BCD:

$$\tan 47°30' = \frac{h}{x} \qquad \text{and} \quad \tan 42°10' = \frac{h}{x + 33}$$

$$h = x \tan 47°30' \qquad\qquad h = (x + 33) \tan 42°10'$$

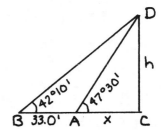

Then we know that h is equal to itself and thus,

$$x \tan 47°30' = (x + 33) \tan 42°10'$$
$$x \tan 47°30' = x \tan 42°10' + 33 \tan 42°10'$$
$$x \tan 47°30' - x \tan 42°10' = 33 \tan 42°10'$$
$$x(\tan 47°30' - \tan 42°10') = 33 \tan 42°10'$$

$$x = \frac{33 \tan 42°10'}{\tan 47°30' - \tan 42°10'}$$

$$x = \frac{33 \tan 42.17°}{\tan 47.5° - \tan 42.17°}$$

$$x = \frac{33(0.9057)}{1.0913 - 0.9057}$$

$$x = \frac{29.8881}{0.1856}$$

$$x = 161 \text{ ft}$$

25. First, we will find the distance, x, each person is from the base of the pole, using the Pythagorean theorem:

$$x^2 + x^2 = (25)^2$$
$$2x^2 = 625$$
$$x^2 = \frac{625}{2}$$
$$x = \frac{25\sqrt{2}}{2} \text{ or } - \frac{25\sqrt{2}}{2}$$
$$x = \frac{25\sqrt{2}}{2}$$

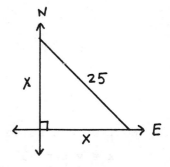

Next, we find the height of the pole, h, using the tangent ratio:

$$\tan 56° = \frac{h}{x}$$

$$h = x \tan 56°$$

$$h = \frac{25\sqrt{2}}{2}(1.4826)$$

$$h = 26 \text{ ft}$$

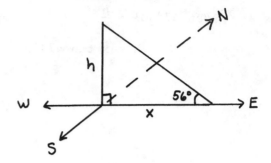

29. We can find x using both $\triangle ADB$ and $\triangle BCD$:

$$\tan 53° = \frac{x}{y} \quad \text{and} \quad \tan 31° = \frac{x}{15 - y}$$

$$x = y \tan 53° \qquad x = (15 - y) \tan 31°$$

$$x = 15 \tan 31° - y \tan 31°$$

Therefore,   x = x

$$y \tan 53° = 15 \tan 31° - y \tan 31°$$
$$y \tan 53° + y \tan 31° = 15 \tan 31°$$
$$y(\tan 53° + \tan 31°) = 15 \tan 31°$$

$$y = \frac{15 \tan 31°}{\tan 53° + \tan 31°}$$

$$y = \frac{15(0.6009)}{1.3270 + 0.6009}$$

$$y = \frac{9.0135}{1.9279}$$

$$y = 4.7 \text{ ft}$$

Now, we can solve for x:

$$x = y \tan 53°$$
$$x = (4.7)(1.3270)$$
$$x = 6.2 \text{ ft}$$

33. $\sin \theta \cot \theta = \sin \theta \cdot \frac{\cos \theta}{\sin \theta}$ Ratio identity

$\qquad = \frac{\sin \theta \cos \theta}{\sin \theta}$ Multiplication of fractions

$\qquad = \cos \theta$ Division of like factors

37. $\sec \theta - \cos \theta = \dfrac{1}{\cos \theta} - \cos \theta$ Reciprocal identity

            $= \dfrac{1}{\cos \theta} - \cos \theta \cdot \dfrac{\cos \theta}{\cos \theta}$ LCD is $\cos \theta$

            $= \dfrac{1 - \cos^2 \theta}{\cos \theta}$ Subtraction of fractions

            $= \dfrac{\sin^2 \theta}{\cos \theta}$ Pythagorean identity

Problem Set 2.5

13. To find the magnitude of $\vec{v}$, we use the Pythagorean theorem:

$$|\vec{v}| = \sqrt{(3.25)^2 + (12.0)^2}$$
$$= \sqrt{10.5625 + 144}$$
$$= \sqrt{154.5625}$$
$$= 12.4 \text{ mph}$$

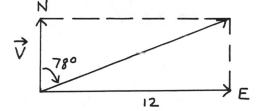

To find the direction of the boat, we must first find θ using the tangent ratio:

$$\tan \theta = \frac{12.0}{3.25}$$
$$\tan \theta = 3.6923$$
$$\theta = 74.8°$$

Therefore, the true course of the boat is 12.4 miles per hour at N 74.8° E.

17. We want to find the magnitude of vector, v, which we can do using the tangent ratio:

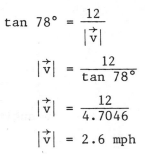

$$\tan 78° = \frac{12}{|\vec{v}|}$$
$$|\vec{v}| = \frac{12}{\tan 78°}$$
$$|\vec{v}| = \frac{12}{4.7046}$$
$$|\vec{v}| = 2.6 \text{ mph}$$

21. To find the distance, x, the plane has flown off its course, we can use the sine ratio:

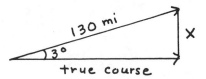

$$\sin 3° = \frac{x}{130}$$
$$x = 130 \sin 3°$$
$$x = 6.8 \text{ miles}$$

25. $|\vec{v}_x| = |\vec{v}| \cos \theta$ $|\vec{v}_y| = |\vec{v}| \sin \theta$

   $= 420 \cos 36°10'$ $= 420 \sin 36°10'$

   $= 420 \cos 36.17°$ $= 420 \sin 36.17°$

   $= 420(0.8073)$ $= 420(0.5901)$

   $= 339$ $= 248$

29. $|\vec{v}| = \sqrt{|\vec{v}_x|^2 + |\vec{v}_y|^2}$

   $= \sqrt{(35.0)^2 + (26.0)^2}$

   $= \sqrt{1225 + 676}$

   $= \sqrt{1901}$

   $= 43.6$

33. $|\vec{v}_x| = |\vec{v}| \cos \theta$ $|\vec{v}_y| = |\vec{v}| \sin \theta$

   $= 1200 \cos 45°$ $= 1200 \sin 45°$

   $= 1200(0.7071)$ $= 1200(0.7071)$

   $= 850$ feet per second $= 850$ feet per second

37. We are given that $|\vec{v}_x| = 35.0$ and $|\vec{v}_y| = 15.0$.

   $|\vec{v}| = \sqrt{|\vec{v}_x|^2 + |\vec{v}_y|^2}$

   $= \sqrt{(35.0)^2 + (15.0)^2}$

   $= \sqrt{1225 + 225}$

   $= \sqrt{1450}$

   $= 38.1$ feet per second

   $\tan \theta = \dfrac{|\vec{v}_y|}{|\vec{v}_x|}$

   $\tan \theta = \dfrac{15.0}{35.0}$

   $\tan \theta = 0.4286$

   $\theta = 23.2°$

Therefore, the velocity of the arrow is 38.1 feet per second at an angle of inclination of 23.2°.

41. To find the total distance traveled north, we must find the sum of $|\vec{v}_y|$ and $|\vec{w}_y|$ and to find the total distance traveled east, we must find the sum of $|\vec{v}_x|$ and $|\vec{w}_x|$.

We are given that $|\vec{v}|$ is 175 mi. at an angle of elevation of 90° - 18° or 72° and also that $|\vec{w}|$ is 120 mi. at an angle of elevation of 90° - 49° or 41°.

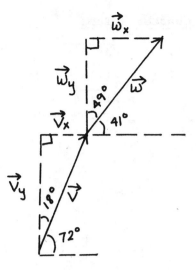

$|\vec{v}_x| = |\vec{v}| \cos \theta_1$
$= 175 \cos 72°$
$= 175(0.3090)$
$= 54$ mi

$|\vec{v}_y| = |\vec{v}| \sin \theta_1$
$= 175 \sin 72°$
$= 175(0.9511)$
$= 166$ mi

$|\vec{w}_x| = |\vec{w}| \cos \theta_2$
$= 120 \cos 41°$
$= 120(0.7547)$
$= 91$ mi

$|\vec{w}_y| = |\vec{w}| \sin \theta_2$
$= 120 \sin 41°$
$= 120(0.6561)$
$= 79$ mi

Therefore, the total distance north is
$$|\vec{v}_y| + |\vec{w}_y| = 166 + 79 = 245 \text{ miles}$$

and the total distance east is
$$|\vec{v}_x| + |\vec{w}_x| = 54 + 91 = 145 \text{ miles}$$

45. A point on the terminal side of θ is (1, 2).

$r = \sqrt{x^2 + y^2}$ where $x = 1$ and $y = 2$

$r = \sqrt{1^2 + 2^2}$

$r = \sqrt{1 + 4}$

$r = \sqrt{5}$

$\sin \theta = \dfrac{y}{r}$ $\cos \theta = \dfrac{1}{\sqrt{5}}$

     $= \dfrac{2}{\sqrt{5}}$

42

1. $c = \sqrt{a^2 + b^2}$

   $= \sqrt{(1)^2 + (2)^2}$

   $= \sqrt{1 + 4}$

   $= \sqrt{5}$

   $\sin A = \dfrac{a}{c}$ $\sin B = \dfrac{b}{c}$

   $\quad = \dfrac{1}{\sqrt{5}}$ $\quad = \dfrac{2}{\sqrt{5}}$

   $\cos A = \dfrac{b}{c}$ $\cos B = \dfrac{a}{c}$

   $\quad = \dfrac{2}{\sqrt{5}}$ $\quad = \dfrac{1}{\sqrt{5}}$

   $\tan A = \dfrac{a}{b}$ $\tan B = \dfrac{b}{a}$

   $\quad = \dfrac{1}{2}$ $\quad = \dfrac{2}{1} = 2$

5. $\sin 14° = \cos (90° - 14°)$

   $\qquad\quad = \cos 76°$

9. $\sin^2 60° - \cos^2 30° = (\frac{\sqrt{3}}{2})^2 - (\frac{\sqrt{3}}{2})^2$

   $\qquad\qquad\qquad\quad = 0$

13. $73.2° = 73° + .2(60)'$

    $\qquad\quad = 73° \ 12'$

17. $\sin 24°20' = \sin 24.33°$

    $\qquad\qquad\; = 0.4120$

29. $A = 90° - 24.9°$

    $\quad = 65.1°$

    $\sin 24.9° = \dfrac{305}{c}$ $\tan 65.1° = \dfrac{a}{305}$

    $\qquad c = \dfrac{305}{\sin 24.9°}$ $a = 305 \tan 65.1°$

    $\qquad c = \dfrac{305}{0.4210}$ $a = 305(2.1543)$

    $\qquad c = 724$ $a = 657$

33. $\tan 43° = \dfrac{35}{x}$

$\qquad x = \dfrac{35}{\tan 43°}$

$\qquad x = \dfrac{35}{0.9325}$

$\qquad x = 37.5$ ft

$\tan 47° = \dfrac{35}{y}$

$\qquad y = \dfrac{35}{\tan 47°}$

$\qquad y = \dfrac{35}{1.0724}$

$\qquad y = 32.6$ ft

$x + y = 70$ feet (rounded to 2 significant digits)

37. $\theta = -30° + 360°$
$\quad = 330°$

$|\vec{v}_x| = |\vec{v}|\cos\theta$
$\qquad = 120\cos 330°$
$\qquad = 120(0.8660)$
$\qquad = 104$

$|\vec{v}_y| = |\vec{v}|\sin\theta$
$\qquad = 120\sin 330°$
$\qquad = 120(-0.5)$
$\qquad = -60$

The ship travels 104 miles east and 60 miles south.

RADIAN MEASURE

<u>Problem Set 3.1</u>

1. 210° − 180° = 30°

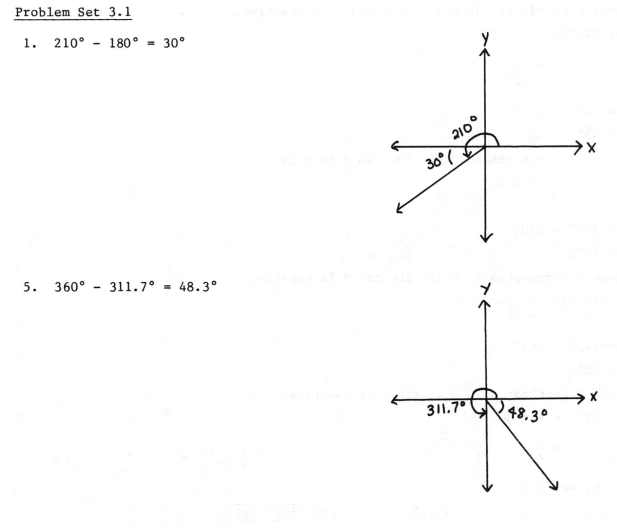

5. 360° − 311.7° = 48.3°

9. −300° + 360° = 60°

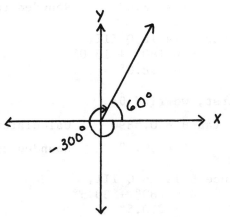

13. $\hat{\theta} = 225° - 180°$

$\hat{\theta} = 45°$

Since θ terminates in Q III, the cos θ is negative.

cos 225° = −cos 45°

$$= -\frac{1}{\sqrt{2}}$$

17. $\hat{\theta} = 180° - 135°$

$\hat{\theta} = 45°$

Since θ terminates in Q II, the tan θ is negative.

tan 135° = −tan 45°
= −1

21. $\hat{\theta} = 360° - 330°$

$\hat{\theta} = 30°$

Since θ terminates in Q IV, the csc θ is negative

csc 330° = −csc 30°
= −2

25. $\hat{\theta} = 390° - 360°$

$\hat{\theta} = 30°$

Since θ terminates in Q I, sin θ is positive.

sin 390° = sin 30°

$$= \frac{1}{2}$$

49. First, we find $\hat{\theta}$:

sin $\hat{\theta}$ = 0.3090   Calculator: 0.3090 ⌊inv⌋ ⌊sin⌋

   $\hat{\theta}$ = 18.0°   Rounded to the nearest tenth of a degree

Since θ is in Q III,
   θ = 180° + 18.0°
   = 198.0°

53. First, we find $\hat{\theta}$:

tan $\hat{\theta}$ = 0.5890   Calculator: 0.5890 ⌊inv⌋ ⌊tan⌋

   $\hat{\theta}$ = 30.5°   Rounded to the nearest tenth of a degree

Since θ is in Q III,
   θ = 180° + 30.5°
   = 210.5°

57. First, we find $\hat{\theta}$:

  $\sin \hat{\theta} = 0.9652$ Calculator: 0.9652 [inv] [sin]

  $\hat{\theta} = 74.8°$ Rounded to the nearest tenth of a degree

Since θ is in Q II,
  $\theta = 180° - 74.8°$
  $= 105.2°$

61. First, we find $\hat{\theta}$:

  $\csc \hat{\theta} = 2.4957$

  $\sin \hat{\theta} = \dfrac{1}{2.4957}$ Calculator: 2.4957 [1/x] [inv] [sin]

  $\hat{\theta} = 23.6°$ Rounded to the nearest tenth of a degree

Since θ is in Q II,
  $\theta = 180° - 23.6°$
  $= 156.4°$

65. First, we find $\hat{\theta}$:

  $\sec \hat{\theta} = 1.7876$

  $\cos \hat{\theta} = \dfrac{1}{1.7876}$ Calculator: 1.7876 [1/x] [inv] [cos]

  $\hat{\theta} = 56.0°$ Rounded to the nearest tenth of a degree

Since θ is in Q III,
  $\theta = 180° + 56.0°$
  $= 236.0°$

69. First, we find $\hat{\theta}$:

  $\cos \hat{\theta} = \dfrac{1}{\sqrt{2}}$ This is an exact value

  $\hat{\theta} = 45°$

Since θ is in Q II,
  $\theta = 180° - 45°$
  $= 135°$

73. First, we find $\hat{\theta}$:

  $\tan \hat{\theta} = \sqrt{3}$ This is an exact value
  $\hat{\theta} = 60°$

Since θ is in Q III,
  $\theta = 180° + 60°$
  $= 240°$

77. First, we find $\hat{\theta}$:

$$\csc \hat{\theta} = \sqrt{2}$$

$$\sin \hat{\theta} = \frac{1}{\sqrt{2}} \qquad \text{This is an exact value}$$

$$\hat{\theta} = 45°$$

Since θ is in Q II,
$$\theta = 180° - 45°$$
$$= 135°$$

81. The complement of 70° is 20° because 70° + 20° = 90°.
The supplement of 70° is 110° because 70° + 110° = 180°.

85. The side opposite the 30° angle is one-half of the longest side, or $\frac{1}{2} \cdot 10 = 5$.

The side opposite the 60° angle is $\sqrt{3}$ times the shortest side, or $5\sqrt{3}$.

89. $\sin^2 45° + \cos^2 45° = (\frac{1}{\sqrt{2}})^2 + (\frac{1}{\sqrt{2}})^2$

$$= \frac{1}{2} + \frac{1}{2}$$

$$= 1$$

Problem Set 3.2

1. $\theta = \dfrac{s}{r}$ Definition of radian measure

$= \dfrac{9 \text{ cm}}{3 \text{ cm}}$ Substitute given values

$= 3$ radians   Divide

5. $\theta = \dfrac{s}{r}$ Definition of radian measure

$= \dfrac{12\pi \text{ inches}}{4 \text{ inches}}$ Substitute given values

$= 3\pi$ radians   Divide

9. $\theta = \dfrac{s}{r}$ Definition of radian measure

$= \dfrac{450}{4000}$ Substitute given values

$= \dfrac{9}{80}$ or .1125 radians   Divide

13. (b) $90° = 90 \cdot \left(\frac{\pi}{180}\right)$

        $= \frac{\pi}{2}$ radians

    (c) Reference angle is itself:

        $90° = \frac{\pi}{2}$ radians

(a)

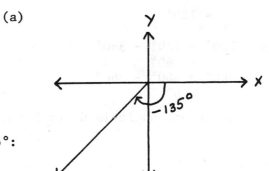

17. (b) $-150° = -150 \left(\frac{\pi}{180}\right)$

        $= -\frac{5\pi}{6}$

    (c) $-150° = -150° + 360°$
        $-150° = 210°$

    We find the reference angle of $210°$:
        $210° - 180° = 30°$

        $30° = 30 \left(\frac{\pi}{180}\right)$

           $= \frac{\pi}{6}$ radians

(a)

21. (b) $-135° = -135 \left(\frac{\pi}{180}\right)$

           $= -\frac{3\pi}{4}$ radians

    (c) $-135° = -135° + 360°$
        $-135° = 225°$

    We find the reference angle of $225°$:

        $225° - 180° = 45°$

        $45° = 45 \left(\frac{\pi}{180}\right)$

           $= \frac{\pi}{4}$ radians

(a)

25. 1 minute $= \frac{1}{60}$ degree

        $= \frac{1}{60} \left(\frac{\pi}{180}\right)$

        $= \frac{\pi}{10,800}$

        $= 0.000290888$

        $= 0.000291$ Rounded to 3 significant digits

29. (a) $\frac{\pi}{3} = \frac{\pi}{3} (\frac{180}{\pi})°$

$= 60°$

(c) Reference angle is itself: $\frac{\pi}{3} = 60°$

(b)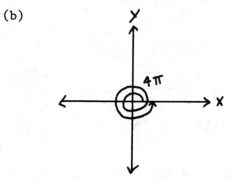

33. (a) $-\frac{7\pi}{6} = -\frac{7\pi}{6} (\frac{180}{\pi})°$

$= -210°$

(c) $-210° = -210° + 360°$
$= 150°$

Reference angle of 150° is
180° - 150° = 30°

$30° = 30 (\frac{\pi}{180})$

$= \frac{\pi}{6}$ radians

(b)

37. (a) $4\pi = 4\pi (\frac{180}{\pi})°$

$= 720°$

(c) $720° = 720° - 360°$
$= 360°$
$360° = 360° - 360°$
$= 0°$

Reference angle is 0° or 0 radians

(b)

41. $1 = 1 (\frac{180}{\pi})°$

$= (\frac{180}{\pi})°$

$= 57.3°$

45. $0.75 = 0.75 (\frac{180}{\pi})°$

$= (\frac{135}{\pi})°$

$= 43.0°$

49. Since $\frac{4\pi}{3}$ terminates in Q III, its sine will be negative.

$\hat{\theta} = \frac{4\pi}{3} - \pi$

$= \frac{\pi}{3}$

$\sin \frac{4\pi}{3} = -\sin \frac{\pi}{3}$ $(\frac{\pi}{3} = 60°)$

$= -\frac{\sqrt{3}}{2}$

53. Since $\frac{2\pi}{3}$ terminates in Q II, its secant will be negative.

$\hat{\theta} = \pi - \frac{2\pi}{3}$

$= \frac{\pi}{3}$

$\sec \frac{2\pi}{3} = -\sec \frac{\pi}{3}$

$= -2$

57. Since $-\frac{\pi}{4}$ terminates in Q IV, its sine will be negative.

$\theta = -\frac{\pi}{4} + 2\pi$

$= \frac{7\pi}{4}$

$\hat{\theta} = 2\pi - \frac{7\pi}{4}$

$= \frac{\pi}{4}$

$4 \sin (-\frac{\pi}{4}) = -4 \sin \frac{\pi}{4}$

$= -4 (\frac{\sqrt{2}}{2})$

$= -2\sqrt{2}$

61. $2 \cos \frac{\pi}{6} = 2 (\frac{\sqrt{3}}{2})$

$= \sqrt{3}$

65. $6 \cos 3 (\frac{\pi}{6}) = 6 \cos \frac{\pi}{2}$

$= 6(0)$

$= 0$

69. $4 \cos \left[2\left(\frac{\pi}{6}\right) + \frac{\pi}{3} \right] = 4 \cos \left(\frac{\pi}{3} + \frac{\pi}{3}\right)$

$$= 4 \cos \frac{2\pi}{3}$$

Since $\frac{2\pi}{3}$ terminates in Q II, its cosine will be negative.

$\hat{\theta} = \pi - \frac{2\pi}{3}$

$\quad = \frac{\pi}{3}$

$4 \cos \frac{2\pi}{3} = -4 \cos \frac{\pi}{3}$

$$= -4 \left(\frac{1}{2}\right)$$

$$= -2$$

73. $(x, y) = (m, n)$

$r = \sqrt{x^2 + y^2}$
$\quad = \sqrt{m^2 + n^2}$

$\sin \theta = \frac{y}{r} = \dfrac{n}{\sqrt{m^2 + n^2}}$ $\qquad \cot \theta = \frac{x}{y} = \frac{m}{n}$

$\cos \theta = \frac{x}{r} = \dfrac{m}{\sqrt{m^2 + n^2}}$ $\qquad \sec \theta = \frac{r}{x} = \dfrac{\sqrt{m^2 + n^2}}{m}$

$\tan \theta = \frac{y}{x} = \frac{n}{m}$ $\qquad \csc \theta = \frac{r}{y} = \dfrac{\sqrt{m^2 + n^2}}{n}$

77. A point on the terminal side, $y = 2x$, in Q I, is $(1, 2)$.

$r = \sqrt{x^2 + y^2}$
$\quad = \sqrt{(1)^2 + (2)^2}$
$\quad = \sqrt{1 + 4}$
$\quad = \sqrt{5}$

$\sin \theta = \frac{y}{r} = \frac{2}{\sqrt{5}}$ $\qquad \cot \theta = \frac{x}{y} = \frac{1}{2}$

$\cos \theta = \frac{x}{r} = \frac{1}{\sqrt{5}}$ $\qquad \sec \theta = \frac{r}{x} = \frac{\sqrt{5}}{1} = \sqrt{5}$

$\tan \theta = \frac{y}{x} = \frac{2}{1} = 2$ $\qquad \csc \theta = \frac{r}{y} = \frac{\sqrt{5}}{2}$

Problem Set 3.3

1. The point on the unit circle is $(-\frac{\sqrt{3}}{2}, \frac{1}{2})$.

$\sin 150° = \frac{1}{2}$

$\cos 150° = -\frac{\sqrt{3}}{2}$

$\tan 150° = \frac{\sin 150°}{\cos 150°} = \frac{1/2}{-\sqrt{3}/2} = -\frac{1}{\sqrt{3}}$

$\cot 150° = \frac{\cos 150°}{\sin 150°} = \frac{-\sqrt{3}/2}{1/2} = -\sqrt{3}$

$\sec 150° = \frac{1}{\cos 150°} = \frac{1}{-\sqrt{3}/2} = -\frac{2}{\sqrt{3}}$

$\csc 150° = \frac{1}{\sin 150°} = \frac{1}{1/2} = 2$

5. The point on the unit circle is $(-1, 0)$.

$\sin 180° = 0$

$\cos 180° = -1$

$\tan 180° = \frac{\sin 180°}{\cos 180°} = \frac{0}{-1} = 0$

$\cot 180° = \frac{\cos 180°}{\sin 180°} = \frac{-1}{0}$ (undefined)

$\sec 180° = \frac{1}{\cos 180°} = \frac{1}{-1} = -1$

$\csc 180° = \frac{1}{\sin 180°} = \frac{1}{0}$ (undefined)

9. $\cos (-60°) = \cos 60°$ cosine is an even function

$= \frac{1}{2}$ from the unit circle

13. $\sin (-30°) = -\sin 30°$ sine is an odd function

$= -\frac{1}{2}$ from the unit circle

17. On the unit circle, we locate all points with a y-coordinate of $\frac{1}{2}$. The angles associated with these points are $\frac{\pi}{6}$ and $\frac{5\pi}{6}$.

21. We look for points on the unit circle where the ratio, $\frac{y}{x}$, equals $-\sqrt{3}$. The angles associated with these points are $\frac{2\pi}{3}$ and $\frac{5\pi}{3}$.

25. The point on the unit circle is (x, y).

    $\sin \theta = y$ and $\cos \theta = x$

    $\tan \theta = \dfrac{\sin \theta}{\cos \theta} = \dfrac{y}{x}$

    $\cot \theta = \dfrac{\cos \theta}{\sin \theta} = \dfrac{x}{y}$

    $\sec \theta = \dfrac{1}{\cos \theta} = \dfrac{1}{x}$

    $\csc \theta = \dfrac{1}{\sin \theta} = \dfrac{1}{y}$

33. $\sin(-\theta)\cot(-\theta) = \sin(-\theta)\,\dfrac{\cos(-\theta)}{\sin(-\theta)}$ Ratio identity

    $= \dfrac{\sin(-\theta)\cos(-\theta)}{\sin(-\theta)}$ Multiplication of fractions

    $= \cos(-\theta)$ Division of like factor

    $= \cos \theta$ Cosine is an even function

37. $\csc \theta + \sin(-\theta) = \csc \theta - \sin \theta$ Sine is an odd function

    $= \dfrac{1}{\sin \theta} - \sin \theta$ Reciprocal identity

    $= \dfrac{1}{\sin \theta} - \sin \theta \cdot \dfrac{\sin \theta}{\sin \theta}$ LCD is $\sin \theta$

    $= \dfrac{1}{\sin \theta} - \dfrac{\sin^2 \theta}{\sin \theta}$ Multiplication of fractions

    $= \dfrac{1 - \sin^2 \theta}{\sin \theta}$ Subtraction of fractions

    $= \dfrac{\cos^2 \theta}{\sin \theta}$ Pythagorean identity

41. $\angle B = 90° - 42°$
       $= 48°$

    $\sin 42° = \dfrac{a}{36}$ $\cos 42° = \dfrac{b}{36}$

    $a = 36 \sin 42°$ $b = 36 \cos 42°$
    $a = 36(0.6691)$ $b = 36(0.7431)$
    $a = 24$ $b = 27$

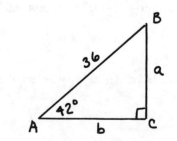

45. $c = \sqrt{a^2 + b^2}$

  $= \sqrt{(20.5)^2 + (31.4)^2}$

  $= \sqrt{420.25 + 985.96}$

  $= \sqrt{1406.21}$

  $= 37.5$

$\tan A = \dfrac{20.5}{31.4}$

$\tan A = 0.6529$

  $A = 33.1°$

$\angle B = 90° - 33.1°$

  $= 56.9°$

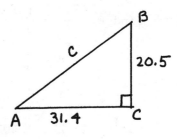

Problem Set 3.4

1. $s = r\,\theta$ Formula for arc length
    $s = 3(2)$ Substitute given values
    $s = 6$ inches   Simplify

5. $s = r\,\theta$ Formula for arc length

    $s = 12\left(\dfrac{\pi}{6}\right)$ Substitute given values

    $s = 2\pi$ centimeters   Simplify

    $s = 6.28$ centimeters   Rounded to 3 significant digits

9. First, we must change θ to radians by multiplying by $\dfrac{\pi}{180}$.

    $s = r\,\theta$ Formula for arc length

    $s = 10(240)\left(\dfrac{\pi}{180}\right)$ Substitute given values (θ in radians)

    $s = \dfrac{40\pi}{3}$ inches   Simplify

    $s = 41.9$ inches   Rounded to 3 significant digits

13. First, we find θ: $\dfrac{\theta}{2\pi} = \dfrac{1}{6}$ One complete rotation is 6 hours or 2π radians

$$\theta = \dfrac{2\pi}{6} \qquad \text{Multiply both sides by } 2\pi$$

$$\theta = \dfrac{\pi}{3} \qquad \text{Simplify}$$

Also, the radius is 200 + 4000 or 4200 miles.

Therefore, $s = r\,\theta$

$$s = 4200\left(\dfrac{\pi}{3}\right)$$

$$s = 1400\pi \text{ miles}$$

$$s = 4400 \text{ miles}$$

17. First we convert $0.5°$ to radians by multiplying by $\dfrac{\pi}{180}$.   Then we apply the formula for arc length.

$$s = r\,\theta$$

$$s = 240,000(0.5)\left(\dfrac{\pi}{180}\right)$$

$$s = \dfrac{2000\pi}{3} \text{ miles}$$

$$s = 2100 \text{ miles}$$

21. $r = \dfrac{s}{\theta}$ Formula for arc length

$r = \dfrac{4.2}{1.4}$ Substitute given values

$r = 3$ inches   Simplify

25. First, we convert θ to radians by multiplying by $\dfrac{\pi}{180}$.

$r = \dfrac{s}{\theta}$ Formula for arc length

$r = \dfrac{\pi/2}{90(\pi/180)}$ Substitute given values (θ in radians)

$r = \dfrac{\pi/2}{\pi/2}$ Simplify

$r = 1$ meter   Simplify

29. $A = \dfrac{1}{2}\,r^2\,\theta$ Formula for area of a sector

$A = \dfrac{1}{2}\,(3)^2(2)$ Substitute given values

$A = 9 \text{ cm}^2$ Simplify

33. $A = \frac{1}{2} r^2 \theta$ Formula for area of a sector

    $A = \frac{1}{2} (3)^2 (\frac{\pi}{5})$ Substitute given values

    $A = \frac{9\pi}{10}$ m^2 Simplify

    $A = 2.83$ m^2 Rounded to 3 significant digits

37. First, we find the radius:

    $r = \frac{s}{\theta}$ Formula for arc length

    $r = \frac{4}{2}$ Substitute given values

    $r = 2$ inches   Simplify

Then, we find the area of the sector:

    $A = \frac{1}{2} r^2 \theta$ Formula for area of a sector

    $A = \frac{1}{2} (2)^2 (2)$ Substitute given values

    $A = 4$ in.^{2} Simplify

41. First we convert θ to radians by multiplying by $\frac{\pi}{180}$.

    $\theta = 45(\frac{\pi}{180})$

    $\theta = \frac{\pi}{4}$

Then we apply the formula for the area of a sector.

    $A = \frac{1}{2} r^2 \theta$ Formula for area of a sector

    $\frac{2\pi}{3} = \frac{1}{2} r^2 (\frac{\pi}{4})$ Substitute given values

    $r^2 = \frac{16}{3}$ Multiply both sides by $\frac{8}{\pi}$

    $r = \frac{4}{\sqrt{3}}$ inches   Take square root of both sides (radius must be positive)

    $r = 2.31$ inches   Rounded to 3 significant digits

45. $\tan \theta = \frac{75}{43}$

$\tan \theta = 1.7442$
$\theta = 60°$

75 ft

43 ft

49. $\tan \theta = \frac{35.8}{10.25}$

$\tan \theta = 3.4927$
$\theta = 74.0°$

35.9 cm

10.25 cm   10.25 cm

20.5 cm

Problem Set 3.5

1. $v = \frac{s}{t}$ Formula for linear velocity

$v = \frac{3}{2}$ Substitute given values

$v = 1.5$ feet per minute   Simplify

5. $v = \frac{s}{t}$ Formula for linear velocity

$v = \frac{30}{2}$ Substitute given values

$v = 15$ miles per hour   Simplify

9. $s = vt$ Formula for linear velocity

$s = 45\left(\frac{1}{2}\right)$ Substitute given values

$s = 22.5$ miles   Simplify

13. $\omega = \frac{\theta}{t}$ Formula for angular velocity

$\omega = \frac{2\pi/3}{5}$ Substitute given values

$\omega = \frac{2\pi}{15}$ radians per second   Simplify

$\omega = .419$ radians per second   Rounded to 3 significant digits

17. $\omega = \frac{\theta}{t}$ Formula for angular velocity

$\omega = \frac{8\pi}{3\pi}$ Substitute given values

$\omega = \frac{8}{3}$ radians per second   Simplify

$\omega = 2.67$ radians per second   Rounded to 3 significant digits

21. We know that $\tan \theta = \frac{d}{100}$.

Therefore, $d = 100 \tan \theta$.

Next, we must find θ. We know that $\omega = \frac{\theta}{t}$. Therefore, $\theta = \omega t$. We also know that $\omega = \frac{2\pi \text{ radians}}{4 \text{ seconds}} = \frac{1}{2}\pi$ radians per second.

We have $\theta = \frac{1}{2}\pi t$.

Substituting θ into $d = 100 \tan \theta$, we get

$d = 100 \tan \frac{1}{2}\pi t$

When $t = \frac{1}{2}$, then $d = 100 \tan \frac{1}{2}(\pi)(\frac{1}{2})$

$$d = 100 \tan \frac{\pi}{4}$$

$$d = 100(1)$$

$$d = 100 \text{ feet}$$

When $t = \frac{3}{2}$, $d = 100 \tan \frac{1}{2}(\pi)(\frac{3}{2})$

$$d = 100 \tan \frac{3\pi}{4}$$

$$d = 100(-1)$$

$$d = -100 \text{ feet}$$

When $t = 1$, $d = 100 \tan \frac{1}{2}(\pi)(1)$

$$d = 100 \tan \frac{\pi}{2}$$

d is undefined because the $\tan \frac{\pi}{2}$ is undefined.

When $t = 1$, $\theta = \frac{\pi}{2}$ and the light rays are parallel to the wall.

25. First we find θ, using the formula for angular velocity. Then we apply the formula for arc length.

$\theta = \omega t$ Formula for angular velocity

$\theta = \frac{3\pi}{2}(30)$ Substitute given values

$\theta = 45\pi$ Simplify

$s = r\theta$ Formula for arc length
$s = 4(45\pi)$ Substitute given values

$s = 180\pi$ meters   Simplify

$s = 565$ meters   Rounded to 3 significant digits

In 29 and 33 we convert revolutions per minute to radians per minute by multiplying by 2π.

29. $\omega = 10(2\pi)$

     $\omega = 20\pi$ radians per minute

     $\omega = 62.8$ radians per minute

33. $\omega = 5.8(2\pi)$

     $\omega = 11.6\pi$ radians per minute

     $\omega = 36.4$ radians per minute

37. Using the relationship between angular velocity and linear velocity, we have

$$\omega = \frac{v}{r}$$

$$\omega = \frac{3}{6}$$

$$\omega = 0.5 \text{ radians per second}$$

41. Angular velocity, ω, is $\frac{1}{24}$ revolutions per hour. To convert this to radians per hour, we multiply by 2π.

$$\omega = \frac{1}{24}(2\pi)$$

$$\omega = \frac{\pi}{12} \text{ radians per hour}$$

$$\omega = 0.262 \text{ radians per hour}$$

45. Using the relationship between angular velocity and linear velocity, we have

$$\omega = \frac{v}{r}$$

$$\omega = \frac{1,100 \text{ ft/sec}}{1 \text{ ft}}$$

$$\omega = 1,100 \text{ radians per second}$$

Then we convert this to revolutions per second by dividing by 2π.

$$\omega = \frac{1100}{2\pi}$$

$$\omega = \frac{550}{\pi} \text{ revolutions per second}$$

To convert this to revolutions per minute, we multiply by 60.

$$\omega = \frac{550}{\pi}(60)$$

$$\omega = \frac{33,000}{\pi} \text{ rpm}$$

$$\omega = 10,500 \text{ rpm}$$

$$= \frac{s}{t}$$ Formula for linear velocity

$$v = \frac{16 \text{ km}}{1 \text{ hr}}$$ Substitute given values

$v = 16$ km per hour   Simplify

$v = 16,000$ meters per hour   Change to meters per hour by multiplying by 1000

$$\omega = \frac{v}{r}$$ Relationship between angular velocity and linear velocity

$$\omega = \frac{16,000 \text{ m/hr}}{.3 \text{ m}}$$ Substitute given values and change radius to meters

$\omega = 53,300$ radians per hour   Simplify

53. $|\vec{v}| = \sqrt{(45.5)^2 + (176)^2}$ Pythagorean theorem

$\quad = \sqrt{2070.25 + 30,976}$ Simplify

$\quad = \sqrt{33,046.25}$ Simplify

$\quad = 182$ mph   Rounded to 3 significant digits

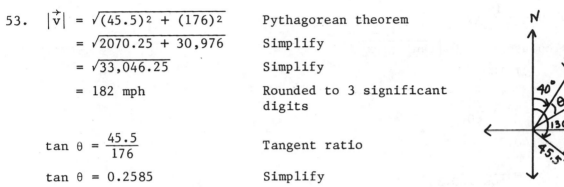

$\tan \theta = \dfrac{45.5}{176}$ Tangent ratio

$\tan \theta = 0.2585$ Simplify

$\theta = 14.5°$ Rounded to the nearest tenth of a degree

The ground speed is 182 mph at 54.5° from due north.

57. $|\vec{v}_x| = 85.5 \cos 32.7°$
$\quad\quad = 85.5(0.8415)$
$\quad\quad = 71.9$

$|\vec{v}_y| = 85.5 \sin 32.7°$
$\quad\quad = 85.5(0.5402)$
$\quad\quad = 46.2$

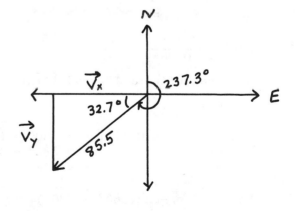

The ship has sailed 71.9 miles west and 46.2 miles south.

Chapter 3 Test

1. $\hat{\theta} = 235° - 180°$
    $= 55°$

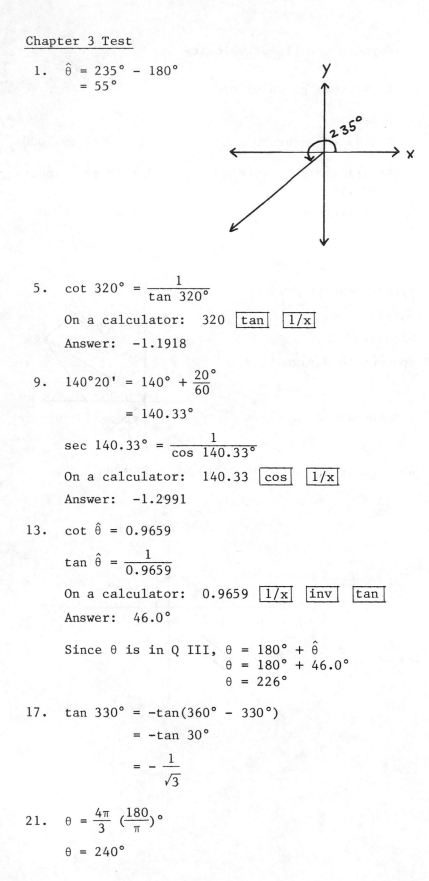

5. $\cot 320° = \dfrac{1}{\tan 320°}$

    On a calculator: 320 $\boxed{\tan}$ $\boxed{1/x}$

    Answer: -1.1918

9. $140°20' = 140° + \dfrac{20°}{60}$

    $= 140.33°$

    $\sec 140.33° = \dfrac{1}{\cos 140.33°}$

    On a calculator: 140.33 $\boxed{\cos}$ $\boxed{1/x}$

    Answer: -1.2991

13. $\cot \hat{\theta} = 0.9659$

    $\tan \hat{\theta} = \dfrac{1}{0.9659}$

    On a calculator: 0.9659 $\boxed{1/x}$ $\boxed{\text{inv}}$ $\boxed{\tan}$

    Answer: 46.0°

    Since θ is in Q III, $\theta = 180° + \hat{\theta}$
    $\theta = 180° + 46.0°$
    $\theta = 226°$

17. $\tan 330° = -\tan(360° - 330°)$

    $= -\tan 30°$

    $= -\dfrac{1}{\sqrt{3}}$

21. $\theta = \dfrac{4\pi}{3} \left(\dfrac{180}{\pi}\right)°$

    $\theta = 240°$

25. $4 \cos \left(-\dfrac{3\pi}{4}\right) = 4 \cos \dfrac{3\pi}{4}$ cosine is an even function

$\quad\qquad = 4\left(-\cos \dfrac{\pi}{4}\right)$ θ is in Q II

$\quad\qquad = 4\left(-\dfrac{1}{\sqrt{2}}\right)$ Substitute exact value

$\quad\qquad = -\dfrac{4}{\sqrt{2}}$ Simplify

$\quad\qquad = -2\sqrt{2}$ Rationalize the denominator

29. $2 \cos \left(3x - \dfrac{\pi}{2}\right) = 2 \cos \left[3\left(\dfrac{\pi}{3}\right) - \dfrac{\pi}{2}\right]$

$\quad\qquad = 2 \cos \left(\pi - \dfrac{\pi}{2}\right)$

$\quad\qquad = 2 \cos \dfrac{\pi}{2}$

$\quad\qquad = 2(0)$

$\quad\qquad = 0$

33. $s = r\,\theta$

$\quad = 12\left(\dfrac{\pi}{6}\right)$

$\quad = 2\pi$ meters

$\quad = 6.28$ meters

37. $A = \dfrac{1}{2} r^2\, \theta$ where $\theta = 90\left(\dfrac{\pi}{180}\right)$ radians

$\qquad\qquad\qquad\qquad\quad \theta = \dfrac{\pi}{2}$ radians

$\quad = \dfrac{1}{2}(4)^2\left(\dfrac{\pi}{2}\right)$

$\quad = 4\pi$ inches^2

$\quad = 12.6$ inches^2

41. $s = vt$
$\quad = 30(3)$
$\quad = 90$ feet

45. $\omega = 6(2\pi)$
$\quad = 12\pi$ radians per minute
$\quad = 37.7$ radians per minute

49. $\omega = 20(2\pi)$
$\quad = 40\pi$ radians per minute

$v = r\omega$
$\quad = 2(40\pi)$
$\quad = 80\pi$ feet per minute
$\quad = 251$ feet per minute

GRAPHING AND INVERSE FUNCTIONS

Problem Set 4.1

9. Refer to the graph in problem 3. Find the x-value corresponding to y = -1.

13. Refer to the graph in problem 5. Find the x-values corresponding to y = 1.

25. Refer to the graph in problem 19. Find the x-values corresponding to y = -1.

29. Refer to the graph in problem 21. Find the x-values corresponding to y = 1.

33. Refer to the graphs in problems 1 and 18. Find the x-values corresponding to y = 0. You will notice the pattern that exists: $x = \frac{\pi}{2} + k\pi$ where k is any integer.

37. Refer to the graphs in problems 5 and 22. Find the x-values corresponding to y = 0. You will notice the pattern that exists: $x = k\pi$ where k is any integer.

41. The amplitude is 2 because the greatest y-value is 2 and the least y-value is -2. The period is 2 because the graph repeats itself every 2 units, or f(x + 2) = f(x).

49. $\cos \theta \tan \theta = \cos \theta \cdot \dfrac{\sin \theta}{\cos \theta}$ Ratio identity

         $= \dfrac{\cos \theta \sin \theta}{\cos \theta}$ Multiplication of fractions

         $= \sin \theta$ Division of like factor

53. $\csc \theta + \sin(-\theta) = \csc \theta - \sin \theta$ Sine is an odd function

         $= \dfrac{1}{\sin \theta} - \sin \theta$ Reciprocal identity

         $= \dfrac{1}{\sin \theta} - \sin \theta \cdot \dfrac{\sin \theta}{\sin \theta}$ LCD is $\sin \theta$

         $= \dfrac{1}{\sin \theta} - \dfrac{\sin^2 \theta}{\sin \theta}$ Multiplication of fractions

         $= \dfrac{1 - \sin^2 \theta}{\sin \theta}$ Subtraction of fractions

         $= \dfrac{\cos^2 \theta}{\sin \theta}$ Pythagorean identity

57. $\dfrac{\pi}{4} = \dfrac{\pi}{4}\left(\dfrac{180}{\pi}\right)^{\circ}$

     $= 45^{\circ}$

61. $\dfrac{11\pi}{6} = \dfrac{11\pi}{6}\left(\dfrac{180}{\pi}\right)^{\circ}$

     $= 330^{\circ}$

Problem Set 4.2

1. $y = 6 \sin x$ Amplitude = 6

    Period = $\frac{2\pi}{1} = 2\pi$

5. $y = \cos \frac{1}{3} x$ Amplitude = 1

    Period = $\frac{2\pi}{1/3} = 6\pi$

9. $y = \sin \pi x$ Amplitude = 1

    Period = $\frac{2\pi}{\pi} = 2$

13. $y = 4 \sin 2x$ Amplitude = 4

    Period = $\frac{2\pi}{2} = \pi$

17. $y = 3 \sin \frac{1}{2} x$ Amplitude = 3

    Period = $\frac{2\pi}{1/2} = 4\pi$

21. $y = \frac{1}{2} \sin \frac{\pi}{2} x$ Amplitude = $\frac{1}{2}$

    Period = $\frac{2\pi}{\pi/2} = 4$

25. $y = 3 \sin 2x$ Amplitude = 3

    Period = $\frac{2\pi}{2} = \pi$

29. $y = -2 \sin (-3x)$
     $y = -2 [-\sin 3x]$ Sine is an odd function
     $y = 2 \sin 3x$ Simplify

    Amplitude = 2

    Period = $\frac{2\pi}{3}$

33. $y = \csc 3x$ Range: $y \geqslant 1$ and $y \leqslant -1$

    Period = $\frac{2\pi}{3}$

Since $y = \csc 3x$ is the reciprocal of $y = \sin 3x$, we sketch $y = \sin 3x$. We draw
a vertical asymptote wherever $y = \sin 3x$ is 0. Using the graph of $y = \sin 3x$
and the vertical asymptotes, we sketch the graph of $y = \csc 3x$.

37. $y = -2 \csc 3x$ Range: $y \geqslant |-2|$ and $y \leqslant -|-2|$
                          $y \geqslant 2$ and $y \leqslant -2$

    Period = $\frac{2\pi}{3}$

    Graph will be reflected about the x-axis

Using the graph from problem 33 as a guide, we change the range and reflect the
graph about the x-axis.

41. $y = \tan 2x$

Period is $\frac{\pi}{2}$. The graph will go through 2 complete cycles every π units.

45. $y = \tan 3x$

Period is $\frac{\pi}{3}$.

One asymptote will be at $\frac{1}{2} \cdot \frac{\pi}{3} = \frac{\pi}{6}$. In general, the asymptotes will be at $\frac{\pi}{6} + \frac{k\pi}{3}$ where k is an integer.

49. $y = -\cot 2x$ Period is $\frac{\pi}{2}$.

Graph is reflected about x-axis

One asymptote will be at 0. In general, the asymptotes will be at $0 + \frac{k\pi}{2}$ where k is an integer.

53. $\sin (x + \frac{\pi}{2}) = \sin (\frac{\pi}{2} + \frac{\pi}{2})$ Substitute $\frac{\pi}{2}$ for x

$= \sin \pi$ Simplify

$= 0$ Exact value

57. $\sin (x + y) = \sin (\frac{\pi}{2} + \frac{\pi}{6})$ Substitute $\frac{\pi}{2}$ for x and $\frac{\pi}{6}$ for y

$= \sin (\frac{2\pi}{3})$ Simplify

$= \sin \frac{\pi}{3}$ Sine is positive in Q II

$= \frac{\sqrt{3}}{2}$ Exact value

61. $45° = 45(\frac{\pi}{180})$ 65. $150° = 150(\frac{\pi}{180})$

$= \frac{\pi}{4}$ $= \frac{5\pi}{6}$

Problem Set 4.3

13. Amplitude = 1

Period = $\frac{2\pi}{\pi} = 2$

Phase Shift = $\frac{-\pi/2}{\pi} = -\frac{1}{2}$

A = Starting point = $-\dfrac{1}{2}$

E = Ending point = $-\dfrac{1}{2} + 2 = \dfrac{3}{2}$

C = Center point = $\dfrac{1}{2}\left(-\dfrac{1}{2} + \dfrac{3}{2}\right) = \dfrac{1}{2}(1) = \dfrac{1}{2}$

B = Average of starting point and center = $\dfrac{1}{2}\left(-\dfrac{1}{2} + \dfrac{1}{2}\right)$

$$= \dfrac{1}{2}(0)$$

$$= 0$$

D = Average of center and ending point = $\dfrac{1}{2}\left(\dfrac{1}{2} + \dfrac{3}{2}\right)$

$$= \dfrac{1}{2}(2)$$

$$= 1$$

The 5 points we use on the x-axis are:

$$-\dfrac{1}{2},\ 0,\ \dfrac{1}{2},\ 1,\ \dfrac{3}{2}$$

The 2 points we use on the y-axis are 1 and -1.

17. Amplitude = 2

Period = $\dfrac{2\pi}{1/2} = 4\pi$

Phase Shift = $\dfrac{-\pi/2}{1/2} = -\pi$

$A = -\pi$

$E = -\pi + 4\pi = 3\pi$

$C = \dfrac{1}{2}(-\pi + 3\pi) = \dfrac{1}{2}(2\pi) = \pi$

$B = \dfrac{1}{2}(-\pi + \pi) = \dfrac{1}{2}(0) = 0$

$D = \dfrac{1}{2}(\pi + 3\pi) = \dfrac{1}{2}(4\pi) = 2\pi$

The 5 points we use on the x-axis are:

$$-\pi,\ 0,\ \pi,\ 2\pi,\ 3\pi$$

The 2 points we use on the y-axis are:

$$2 \text{ and } -2$$

21. Amplitude = 3

Period = $\dfrac{2\pi}{\pi/3}$ = 6

Phase shift = $\dfrac{\pi/3}{\pi/3}$ = 1

A = 1

E = 1 + 6 = 7

C = $\dfrac{1}{2}$(1 + 7) = $\dfrac{1}{2}$(8) = 4

B = $\dfrac{1}{2}$(1 + 4) = $\dfrac{1}{2}$(5) = $\dfrac{5}{2}$

D = $\dfrac{1}{2}$(4 + 7) = $\dfrac{1}{2}$(11) = $\dfrac{11}{2}$

The 5 points we use on the x-axis are:

$1, \dfrac{5}{2}, 4, \dfrac{11}{2}, 7$

The 2 points we use on the y-axis are:

3 and -3

25. Amplitude = $|-4|$ = 4

Period = $\dfrac{2\pi}{2}$ = π

Phase shift = $\dfrac{\pi/2}{2}$ = $\dfrac{\pi}{4}$

Graph is reflected about the x-axis

A = $\dfrac{\pi}{4}$

E = $\dfrac{\pi}{4}$ + π = $\dfrac{5\pi}{4}$

C = $\dfrac{1}{2}$($\dfrac{\pi}{4}$ + $\dfrac{5\pi}{4}$) = $\dfrac{1}{2}$($\dfrac{3\pi}{2}$) = $\dfrac{3\pi}{4}$

B = $\dfrac{1}{2}$($\dfrac{\pi}{4}$ + $\dfrac{3\pi}{4}$) = $\dfrac{1}{2}$(π) = $\dfrac{\pi}{2}$

D = $\dfrac{1}{2}$($\dfrac{3\pi}{4}$ + $\dfrac{5\pi}{4}$) = $\dfrac{1}{2}$(2π) = π

For one complete cycle, the points we use on the x-axis are $\dfrac{\pi}{4}, \dfrac{\pi}{2}, \dfrac{3\pi}{4}, \pi, \dfrac{5\pi}{4}$.

The points we use on the y-axis are 4 and -4. We must reflect our graph about the x-axis and extend it from $-\dfrac{\pi}{4}$ to $\dfrac{3\pi}{2}$.

29. Amplitude $= \left| -\frac{2}{3} \right| = \frac{2}{3}$

Period $= \frac{2\pi}{3}$

Phase shift $= \frac{-\pi/2}{3} = -\frac{\pi}{6}$

Graph is reflected about the x-axis

$A = -\frac{\pi}{6}$

$E = -\frac{\pi}{6} + \frac{2\pi}{3} = \frac{\pi}{2}$

$C = \frac{1}{2}(-\frac{\pi}{6} + \frac{\pi}{2}) = \frac{1}{2}(\frac{\pi}{3}) = \frac{\pi}{6}$

$B = \frac{1}{2}(-\frac{\pi}{6} + \frac{\pi}{6}) = \frac{1}{2}(0) = 0$

$D = \frac{1}{2}(\frac{\pi}{6} + \frac{\pi}{2}) = \frac{1}{2}(\frac{2\pi}{3}) = \frac{\pi}{3}$

For one complete cycle, the points we use on the x-axis are $-\frac{\pi}{6}$, 0, $\frac{\pi}{6}$, $\frac{\pi}{3}$, $\frac{\pi}{2}$.

The points we use on the y-axis are $\frac{2}{3}$ and $-\frac{2}{3}$.

We must reflect our graph about the x-axis and extend it from $-\pi$ to π.

33. First we will sketch the reciprocal function, $y = 2 \cos (2x - \frac{\pi}{2})$.

Amplitude $= 2$

Period $= \frac{2\pi}{2} = \pi$

Phase shift $= \frac{\pi/2}{2} = \frac{\pi}{4}$

$A = \frac{\pi}{4}$

$E = \frac{\pi}{4} + \pi = \frac{5\pi}{4}$

$C = \frac{1}{2}(\frac{\pi}{4} + \frac{5\pi}{4}) = \frac{1}{2}(\frac{3\pi}{2}) = \frac{3\pi}{4}$

$B = \frac{1}{2}(\frac{\pi}{4} + \frac{3\pi}{4}) = \frac{1}{2}(\pi) = \frac{\pi}{2}$

$D = \frac{1}{2}(\frac{3\pi}{4} + \frac{5\pi}{4}) = \frac{1}{2}(2\pi) = \pi$

The 5 points on the x-axis will be:

$$\frac{\pi}{4}, \ \frac{\pi}{2}, \ \frac{3\pi}{4}, \ \pi, \ \frac{5\pi}{4}$$

The 2 points on the y-axis will be 2 and -2.

We sketch this cosine curve using a dotted line.

Then we use the cosine curve to graph $y = 2 \sec (2x - \frac{\pi}{2})$.

The asymptotes will occur where $y = 0$, that is, at $\frac{\pi}{2}$ and π.

Using the asymptotes and the cosine curve, we sketch the graph.

37. Period = π

Phase shift = $-\frac{\pi}{4}$

Asymptotes will occur at $[\frac{\pi}{2} + (-\frac{\pi}{4})] + k\pi$

or $\frac{\pi}{4} + k\pi$ where k is an integer

We will use an interval from $-\frac{3\pi}{4}$ to $-\frac{\pi}{4}$ to sketch the tangent curve.

The x-intercept will be $0 + (-\frac{\pi}{4}) = -\frac{\pi}{4}$ and the asymptotes are $x = -\frac{3\pi}{4}$ and $x = \frac{\pi}{4}$.

41. Period = $\frac{\pi}{2}$

Phase shift = $\frac{\pi/2}{2} = \frac{\pi}{4}$

Asymptotes will occur at $(\frac{\pi/2}{2} + \frac{\pi}{4}) + k(\frac{\pi}{2})$

or $\frac{\pi}{2} + \frac{k\pi}{2}$ where k is an integer

We will use an interval from 0 to $\frac{\pi}{2}$ to sketch the tangent curve.

The x-intercept will be $0 + \frac{\pi}{4} = \frac{\pi}{4}$ and the asymptotes are $x = 0$ and $x = \frac{\pi}{2}$.

45. We use a proportion to find θ:

$$\frac{\theta}{2\pi} = \frac{30}{60}$$

$$\frac{\theta}{2\pi} = \frac{1}{2}$$

$$\theta = \pi$$

$s = r\theta$
$s = 2.6\pi$ centimeters
$s = 8.2$ centimeters

49. $A = \frac{1}{2} r^2 \theta$ where θ is in radians

$= \frac{1}{2}(8)^2 (45 \cdot \frac{\pi}{180})$

$= \frac{1}{2}(64)(\frac{\pi}{4})$

$= 8\pi$ inches^2

$= 25.1$ inches^2

Problem Set 4.4

1. We let $y_1 = 1$ and $y_2 = \sin x$ and graph y_1, y_2, and $y = y_1 + y_2$ on the same coordinate system.

5. We let $y_1 = 4$ and $y_2 = 2 \sin x$ (which is a sine curve with an amplitude of 2 and a period of 2π). Then we graph y_1, y_2, and $y = y_1 + y_2$ on the same coordinate system.

9. We let $y_1 = \frac{1}{2}x$ and $y_2 = -\cos x$ (which is a cosine curve reflected about the x-axis). Then we graph y_1, y_2, and $y = y_1 + y_2$ on the same coordinate system.

13. We let $y_1 = 3 \sin x$ (which is a sine curve with an amplitude of 3 and a period of 2π) and $y_2 = \cos 2x$ (which is a cosine curve with an amplitude of 1 and a period of $\frac{2\pi}{2}$ or π.) Then we graph y_1, y_2, and $y = y_1 + y_2$ on the same coordinate system.

17. We let $y_1 = \sin x$ and $y_2 = \sin \frac{x}{2}$ (which is a sine curve with amplitude of 1 and period of $\frac{2\pi}{1/2}$ or 4π). Then we graph y_1, y_2, and $y = y_1 + y_2$ on the same coordinate system.

21. We let $y_1 = \cos x$ and $y_2 = \frac{1}{2} \sin 2x$ (which is a sine curve with amplitude of $\frac{1}{2}$ and period of $\frac{2\pi}{2}$ or π.) Then we graph y_1, y_2, and $y = y_1 + y_2$ on the same coordinate system.

25. $y = x \sin x$

x	0	$\frac{\pi}{2}$	π	$\frac{3\pi}{2}$	2π	$\frac{5\pi}{2}$	3π	$\frac{7\pi}{2}$	4π
$\sin x$	0	1	0	-1	0	1	0	-1	0
$y = x \sin x$	0	$\frac{\pi}{2}$	0	$-\frac{3\pi}{2}$	0	$\frac{5\pi}{2}$	0	$-\frac{7\pi}{2}$	0

29. $s = vt$

$s = (20 \frac{ft}{sec})(60 \text{ sec})$ We change 1 minute to 60 seconds so that units will agree

$= 1200$ ft.

33. 120 rpm $= (120)(2\pi)$ radians per minute
$= 240\pi$ radians per minute

To convert this to radians per second, we divide by 60 because there are 60 seconds in 1 minute.

240π radians per minute $= \dfrac{240\pi}{60}$

$= 4\pi$ radians per second

Problem Set 4.5

1. We interchange x and y and then solve for y in terms of x:
$x = y^2 + 4$ or $y^2 + 4 = x$
$y^2 = x - 4$
$y = \pm\sqrt{x - 4}$

5. We interchange x and y and then solve for y in terms of x:
$x = 3y - 2$ or $3y - 2 = x$
$3y = x + 2$
$y = \dfrac{x + 2}{3}$

9. We interchange x and y and then solve for y in terms of x:
$x = 3^y$
$y = \log_3 x$

Then we graph these on the same coordinate axes.

$y = 3^x$

x	y
-2	$\frac{1}{9}$
-1	$\frac{1}{3}$
0	1
1	3
2	9

$x = 3^y$

x	y
$\frac{1}{9}$	-2
$\frac{1}{3}$	-1
1	0
3	1
9	2

17. The domain of $y = \cos x$ is all real numbers and its range is all real numbers between -1 and 1, or $-1 \leqslant y \leqslant 1$.

    To find its inverse, we interchange the domain and the range. Therefore, the range of $y = \text{arc } \cos x$ is the domain of $y = \cos x$ or all real numbers.

21. $y = \cos^{-1} \frac{1}{2}$ is equivalent to $\cos y = \frac{1}{2}$.

    This means y is an angle with a cosine of $\frac{1}{2}$.

    Therefore, $y = 60° + 360°k$ or $y = 300° + 360°k$ where k = integer.

25. $y = \arccos(0)$ is equivalent to $\cos y = 0$.
    This means y is an angle with a cosine of 0.
    Therefore, $y = 90° + 360°k$ or $y = 270° + 360°k$ where k is an integer.
    This can be simplified to $y = 90° + 180°k$ where k is an integer.

29. $y = \tan^{-1}(1)$ is equivalent to $\tan y = 1$.
    This means y is an angle with a tangent of 1.
    Therefore, $y = 45° + 180°k$ where k is an integer.

33. $y = \sin^{-1}\left(-\frac{1}{\sqrt{2}}\right)$ is equivalent to $\sin y = -\frac{1}{\sqrt{2}}$.

    This means y is an angle with a sine of $-\frac{1}{\sqrt{2}}$.

    The reference angle is $\frac{\pi}{4}$ and the sine is negative in Q III and Q IV.

    Therefore, $y = \frac{5\pi}{4} + 2k\pi$ or $y = \frac{7\pi}{4} + 2k\pi$.

37. $y = \cos^{-1}\left(\frac{\sqrt{3}}{2}\right)$ is equivalent to $\cos y = \frac{\sqrt{3}}{2}$.

    This means y is an angle with a cosine of $\frac{\sqrt{3}}{2}$.

    The reference angle is $\frac{\pi}{6}$ and the cosine is positive in Q I and Q IV.

    Therefore, $y = \frac{\pi}{6} + 2k\pi$ or $y = \frac{11\pi}{6} + 2k\pi$.

41. $y = \csc^{-1}(1)$ is equivalent to $\csc y = 1$.
    This means y is an angle with a cosecant of 1 and also a sine of 1.

    Therefore, $y = \frac{\pi}{2} + 2k\pi$.

45. $y = \arccos(1)$ is equivalent to $\cos y = 1$.
    This means y is an angle with a cosine of 1.
    Therefore, $y = 0 + 2k\pi$ or simply $y = 2k\pi$.

49. $236°$ is in Q III.
    The reference angle is $236° - 180° = 56°$

53. sin $\hat{\theta}$ = 0.7455

   $\hat{\theta}$ = 48.2°

   Since θ is in Q II, θ = 180° − $\hat{\theta}$

                        = 180° − 48.2°

                        = 131.8°

57. cot $\hat{\theta}$ = 0.2089

   tan $\hat{\theta}$ = $\dfrac{1}{0.2089}$

   $\hat{\theta}$ = 78.2°

   Since θ is in Q IV, θ = 360° − $\hat{\theta}$

                        = 360° − 78.2°

                        = 281.8°

Problem Set 4.6

1. The angle between − $\dfrac{\pi}{2}$ and $\dfrac{\pi}{2}$ whose sine is $\dfrac{\sqrt{3}}{2}$ is $\dfrac{\pi}{3}$.

5. The angle between − $\dfrac{\pi}{2}$ and $\dfrac{\pi}{2}$ whose tangent is 1 is $\dfrac{\pi}{4}$.

9. The angle between − $\dfrac{\pi}{2}$ and $\dfrac{\pi}{2}$ whose sine is − $\dfrac{1}{2}$ is − $\dfrac{\pi}{6}$.

13. The angle between − $\dfrac{\pi}{2}$ and $\dfrac{\pi}{2}$ whose sine is 0 is 0.

17. The angle between 0 and π whose cosine is − $\dfrac{1}{2}$ is $\dfrac{2\pi}{3}$.

21. Calculator: 0.1702 $\boxed{\text{inv}}$ $\boxed{\text{sin}}$
   Answer: 9.8°

25. Calculator: 0.3799 $\boxed{\text{inv}}$ $\boxed{\text{tan}}$
   Answer: 20.8°

29. Calculator: 0.4664 $\boxed{\text{+/−}}$ $\boxed{\text{inv}}$ $\boxed{\text{cos}}$
   Answer: 117.8°

33. Calculator: 0.7660 $\boxed{\text{+/−}}$ $\boxed{\text{inv}}$ $\boxed{\text{sin}}$
   Answer: −50.0°

37. Let $\theta = \text{Sin}^{-1}\ \frac{3}{5}$, the $\sin\theta = \frac{3}{5}$ and $-\frac{\pi}{2} \leqslant \theta \leqslant \frac{\pi}{2}$.

Next, we draw a triangle and find the adjacent side using the Pythagorean theorem.

$$\begin{aligned}
\text{adjacent side} &= \sqrt{5^2 - 3^2}\\
&= \sqrt{25 - 9}\\
&= \sqrt{16}\\
&= 4
\end{aligned}$$

From the figure, we find $\tan\theta = \frac{3}{4}$.

41. Let $\theta = \text{Cos}^{-1}\ \frac{1}{2}$, then $\cos\theta = \frac{1}{2}$ and $0 \leqslant \theta \leqslant \pi$.

Next, we know that $\cos\frac{\pi}{3} = \frac{1}{2}$ and $\theta = \frac{\pi}{3}$.

Now we find $\sin\frac{\pi}{3} = \frac{\sqrt{3}}{2}$.

45. Let $\theta = \text{Sin}^{-1}\ \frac{3}{5}$, then $\sin\theta = \frac{3}{5}$ and $-\frac{\pi}{2} \leqslant \theta \leqslant \frac{\pi}{2}$.

Next, we draw a triangle as in problem 37. (See figure above).

From the figure above, we can see that $\sin\theta = \frac{3}{5}$.

49. Let $\theta = \text{Tan}^{-1}\ \frac{1}{2}$, then $\tan\theta = \frac{1}{2}$ and $-\frac{\pi}{2} \leqslant \theta \leqslant \frac{\pi}{2}$.

We are asked to find $\tan\theta$ which is $\frac{1}{2}$.

53. Let $\theta = \text{Sin}^{-1}\ x$, then $\sin\theta = x$ and $-\frac{\pi}{2} \leqslant \theta \leqslant \frac{\pi}{2}$.

Next, we draw a triangle and label the opposite side and the hypotenuse, as shown at the right.
Now we find the adjacent side using the Pythagorean theorem.

$$\begin{aligned}
\text{Adjacent side} &= \sqrt{1^2 - x^2}\\
&= \sqrt{1 - x^2}
\end{aligned}$$

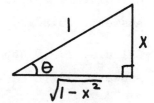

From the figure, we find $\cos\theta = \dfrac{\sqrt{1 - x^2}}{1}$
$$= \sqrt{1 - x^2}$$

57. Let $\theta = \text{Cos}^{-1} \frac{1}{x}$, then $\cos \theta = \frac{1}{x}$ and $0 \leqslant \theta \leqslant \pi$.

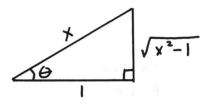

Next, we draw a triangle and label the adjacent side and the hypotenuse.

Now we find the opposite side using the Pythagorean theorem.

$$\text{Opposite side} = \sqrt{x^2 - 1^2}$$
$$= \sqrt{x^2 - 1}$$

From the figure, we find $\sin \theta = \dfrac{\sqrt{x^2 - 1}}{x}$

61. The graph is a sine curve with an amplitude of 4 and a period of $\frac{2\pi}{2}$ or π.

65. The graph is a cosine curve that is reflected about the x-axis with an amplitude of 3 and a period of $\frac{2\pi}{1/2}$ or 4π.

69. Amplitude = 1

Period = $\dfrac{2\pi}{2} = \pi$

Phase Shift = $-\dfrac{\pi/2}{2} = -\dfrac{\pi}{4}$

A (starting point) = $-\dfrac{\pi}{2}$

E (ending point) = $-\dfrac{\pi}{2} + \pi = \dfrac{\pi}{2}$

C (center of above 2 points) = $\dfrac{1}{2}(-\dfrac{\pi}{2} + \dfrac{\pi}{2}) = 0$

B = $\dfrac{1}{2}(-\dfrac{\pi}{2} + 0) = -\dfrac{\pi}{4}$

D = $\dfrac{1}{2}(0 + \dfrac{\pi}{2}) = \dfrac{\pi}{4}$

The 5 points we use on the x-axis are:

$-\dfrac{\pi}{2}, -\dfrac{\pi}{4}, 0, \dfrac{\pi}{4}, \dfrac{\pi}{2}$

The 2 points we use on the y-axis are:

−1 and 1

Chapter 4 Test

9. Amplitude = 1

   Period = $\frac{2\pi}{\pi}$ = 2

13. Amplitude = 1

    Period = 2π

    Phase shift = $-\frac{\pi}{4}$

    $A = -\frac{\pi}{4}$

    $E = -\frac{\pi}{4} + 2\pi = \frac{7\pi}{4}$

    $C = \frac{1}{2}(-\frac{\pi}{4} + \frac{7\pi}{4}) = \frac{1}{2}(\frac{3\pi}{2}) = \frac{3\pi}{4}$

    $B = \frac{1}{2}(-\frac{\pi}{4} + \frac{3\pi}{4}) = \frac{1}{2}(\frac{\pi}{2}) = \frac{\pi}{4}$

    $D = \frac{1}{2}(\frac{3\pi}{4} + \frac{7\pi}{4}) = \frac{1}{2}(\frac{5\pi}{2}) = \frac{5\pi}{4}$

    Points we use on x-axis are $-\frac{\pi}{4}, \frac{\pi}{4}, \frac{3\pi}{4}, \frac{5\pi}{4}, \frac{7\pi}{4}$.

    Points we use on y-axis are −1 and 1.

17. Using problem 13 above, draw in the asymptotes where x = 0, at $-\frac{\pi}{4}, \frac{3\pi}{4}, \frac{7\pi}{4}$.

    Now sketch in the reciprocal function using the asymptotes and the curve

    $y = \sin(x + \frac{\pi}{4})$ as your guides.

21. Let $y_1 = \frac{1}{2}x$ (a straight line) and $y_2 = -\sin x$ (a sine curve reflected about

    the x-axis). Then graph y_1, y_2, and $y = y_1 + y_2$ on the same coordinate system.

29. The angle between $-\frac{\pi}{2}$ and $\frac{\pi}{2}$ whose tangent is −1 is $-\frac{\pi}{4}$.

33. Calculator: 0.6981 $\boxed{+/-}$ $\boxed{\text{inv}}$ $\boxed{\text{cos}}$
    Answer: 134.3°

37. Let $\theta = \text{Cos}^{-1} x$. Then $\cos \theta = x$ and $0 \leqslant \theta \leqslant \pi$.

Next, we draw a triangle and label the adjacent side and the hypotenuse.

Now we find the opposite side using the Pythagorean theorem.

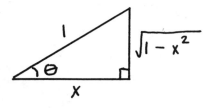

$$\text{Opposite side} = \sqrt{1^2 - x^2}$$
$$= \sqrt{1 - x^2}$$

From the figure, we find $\sin \theta = \dfrac{\sqrt{1 - x^2}}{1}$
$$= \sqrt{1 - x^2}$$

IDENTITIES AND FORMULAS

Problem Set 5.1

1. $\cos \theta \tan \theta = \cos \theta \cdot \dfrac{\sin \theta}{\cos \theta}$ Ratio identity

   $= \dfrac{\cos \theta \sin \theta}{\cos \theta}$ Multiply

   $= \sin \theta$ Reduce

5. $\dfrac{\tan A}{\sec A} = \dfrac{\dfrac{\sin A}{\cos A}}{\dfrac{1}{\cos A}}$ Ratio identity and reciprocal identity

   $= \dfrac{\sin A \cos A}{\cos A}$ Divide

   $= \sin A$ Reduce

9. $\cos x(\csc x + \tan x) = \cos x \csc x + \cos x \tan x$ Distributive property

   $= \cos x \cdot \dfrac{1}{\sin x} + \cos x \cdot \dfrac{\sin x}{\cos x}$ Reciprocal identity and ratio identity

   $= \dfrac{\cos x}{\sin x} + \dfrac{\cos x \sin x}{\cos x}$ Multiply

   $= \cot x + \sin x$ Ratio identity and reduce second fraction

13. $\cos^2 x(1 + \tan^2 x) = \cos^2 x(\sec^2 x)$ Pythagorean identity

   $= \cos^2 x\left(\dfrac{1}{\cos^2 x}\right)$ Reciprocal identity

   $= \dfrac{\cos^2 x}{\cos^2 x}$ Multiply

   $= 1$ Reduce

17. $\dfrac{\cos^4 t - \sin^4 t}{\sin^2 t} = \dfrac{(\cos^2 t + \sin^2 t)(\cos^2 t - \sin^2 t)}{\sin^2 t}$ Factor

   $= \dfrac{1(\cos^2 t - \sin^2 t)}{\sin^2 t}$ Pythagorean identity

   $= \dfrac{\cos^2 t}{\sin^2 t} - \dfrac{\sin^2 t}{\sin^2 t}$ Separate into 2 fractions

   $= \cot^2 t - 1$ Ratio identity and reduce second fraction

21. $\dfrac{1 - \sin^4 \theta}{1 + \sin^2 \theta} = \dfrac{(1 - \sin^2 \theta)(1 + \sin^2 \theta)}{1 + \sin^2 \theta}$ Factor

   $= 1 - \sin^2 \theta$ Reduce

   $= \cos^2 \theta$ Pythagorean identity

25. $\sec^4 \theta - \tan^4 \theta = (\sec^2 \theta - \tan^2 \theta)(\sec^2 \theta + \tan^2 \theta)$ Factor

$\qquad\qquad\qquad = 1(\sec^2 \theta + \tan^2 \theta)$ Pythagorean identity

$\qquad\qquad\qquad = \dfrac{1}{\cos^2 \theta} + \dfrac{\sin^2 \theta}{\cos^2 \theta}$ Reciprocal identity and ratio identity

$\qquad\qquad\qquad = \dfrac{1 + \sin^2 \theta}{\cos^2 \theta}$ Add fractions

29. $\csc B - \sin B = \dfrac{1}{\sin B} - \sin B$ Reciprocal identity

$\qquad\qquad\qquad = \dfrac{1}{\sin B} - \sin B \cdot \dfrac{\sin B}{\sin B}$ LCD is sin B

$\qquad\qquad\qquad = \dfrac{1 - \sin^2 B}{\sin B}$ Subtract fractions

$\qquad\qquad\qquad = \dfrac{\cos^2 B}{\sin B}$ Pythagorean identity

$\qquad\qquad\qquad = \dfrac{\cos B}{\sin B} \cdot \cos B$ Separate fractions

$\qquad\qquad\qquad = \cot B \cos B$ Ratio identity

33. $\dfrac{\cos x}{1 + \sin x} + \dfrac{1 + \sin x}{\cos x} = \dfrac{\cos x}{1 + \sin x} \cdot \dfrac{\cos x}{\cos x} + \dfrac{1 + \sin x}{\cos x} \cdot \dfrac{1 + \sin x}{1 + \sin x}$ LCD

$\qquad\qquad\qquad = \dfrac{\cos^2 x}{\cos x(1 + \sin x)} + \dfrac{1 + 2 \sin x + \sin^2 x}{\cos x(1 + \sin x)}$ Multiply fractions

$\qquad\qquad\qquad = \dfrac{(\cos^2 x + \sin^2 x) + 1 + 2 \sin x}{\cos x(1 + \sin x)}$ Add fractions

$\qquad\qquad\qquad = \dfrac{1 + 1 + 2 \sin x}{\cos x(1 + \sin x)}$ Pythagorean identity

$\qquad\qquad\qquad = \dfrac{2 + 2 \sin x}{\cos x(1 + \sin x)}$ Add

$\qquad\qquad\qquad = \dfrac{2(1 + \sin x)}{\cos x(1 + \sin x)}$ Factor out a 2

$\qquad\qquad\qquad = \dfrac{2}{\cos x}$ Reduce

$\qquad\qquad\qquad = 2 \sec x$ Reciprocal identity

37. $\dfrac{1 - \sec x}{1 + \sec x} = \dfrac{1 - \dfrac{1}{\cos x}}{1 + \dfrac{1}{\cos x}}$ Reciprocal identity

$\qquad\qquad\qquad = \dfrac{\cos x\left(1 - \dfrac{1}{\cos x}\right)}{\cos x\left(1 + \dfrac{1}{\cos x}\right)}$ Multiply numerator and denominator by LCD

$\qquad\qquad\qquad = \dfrac{\cos x - 1}{\cos x + 1}$ Distributive property

41. $\dfrac{1 - \sin t}{1 + \sin t} = \dfrac{1 - \sin t}{1 + \sin t} \cdot \dfrac{1 - \sin t}{1 - \sin t}$ Multiply numerator and denominator by $1 - \sin t$

$\qquad\qquad = \dfrac{(1 - \sin t)^2}{1 - \sin^2 t}$ Multiply fractions

$\qquad\qquad = \dfrac{(1 - \sin t)^2}{\cos^2 t}$ Pythagorean identity

45. $(\sec x - \tan x)^2 = \left(\dfrac{1}{\cos x} - \dfrac{\sin x}{\cos x}\right)^2$ Reciprocal identity and ratio identity

$\qquad\qquad = \left(\dfrac{1 - \sin x}{\cos x}\right)^2$ Subtract fractions

$\qquad\qquad = \dfrac{(1 - \sin x)^2}{\cos^2 x}$ Square fraction

$\qquad\qquad = \dfrac{(1 - \sin x)^2}{(1 - \sin^2 x)}$ Pythagorean identity

$\qquad\qquad = \dfrac{(1 - \sin x)(1 - \sin x)}{(1 - \sin x)(1 + \sin x)}$ Factor

$\qquad\qquad = \dfrac{1 - \sin x}{1 + \sin x}$ Reduce

49. $\dfrac{\sin x + 1}{\cos x + \cot x} = \dfrac{\sin x + 1}{\cos x + \dfrac{\cos x}{\sin x}}$ Ratio identity

$\qquad\qquad = \dfrac{\sin x}{\sin x}\dfrac{(\sin x + 1)}{\left(\cos x + \dfrac{\cos x}{\sin x}\right)}$ Multiply numerator and denominator by LCD

$\qquad\qquad = \dfrac{\sin x(\sin x + 1)}{\sin x \cos x + \cos x}$ Distributive property

$\qquad\qquad = \dfrac{\sin x(\sin x + 1)}{\cos x(\sin x + 1)}$ Factor

$\qquad\qquad = \dfrac{\sin x}{\cos x}$ Reduce

$\qquad\qquad = \tan x$ Ratio identity

53. $\dfrac{\sin^2 B - \tan^2 B}{1 - \sec^2 B} = \dfrac{\sin^2 B - \dfrac{\sin^2 B}{\cos^2 B}}{1 - \dfrac{1}{\cos^2 B}}$ Ratio identity and reciprocal identity

$\qquad\qquad = \dfrac{\cos^2 B\left(\sin^2 B - \dfrac{\sin^2 B}{\cos^2 B}\right)}{\cos^2 B\left(1 - \dfrac{1}{\cos^2 B}\right)}$ Multiply numerator and denominator by LCD

$\qquad\qquad = \dfrac{\cos^2 B \sin^2 B - \sin^2 B}{\cos^2 B - 1}$ Distributive property

$\qquad\qquad = \dfrac{\sin^2 B(\cos^2 B - 1)}{\cos^2 B - 1}$ Factor

$\qquad\qquad = \sin^2 B$ Reduce

57. $\dfrac{\sin^3 A - 8}{\sin A - 2} = \dfrac{(\sin A - 2)(\sin^2 A + 2 \sin A + 4)}{\sin A - 2}$ Factor as difference of 2 cubes

$\quad\quad\quad\quad = \sin^2 A + 2 \sin A + 4$ Reduce

61. $\dfrac{1 - \sin B}{\cos^3 B} = \dfrac{1 - \sin B}{\cos B(\cos^2 B)}$ Factor

$\quad\quad\quad = \dfrac{1 - \sin B}{\cos B(1 - \sin^2 B)}$ Pythagorean identity

$\quad\quad\quad = \dfrac{1 - \sin B}{\cos B(1 + \sin B)(1 - \sin B)}$ Factor

$\quad\quad\quad = \dfrac{1}{\cos B(1 + \sin B)}$ Reduce

$\quad\quad\quad = \dfrac{\sec B}{1 + \sin B}$ Reciprocal identity

65. $\sin(A + B) = \sin(30° + 60°)$

$\quad\quad\quad\quad = \sin 90°$

$\quad\quad\quad\quad = 1$

$\sin A + \sin B = \sin 30° + \sin 60°$

$\quad\quad\quad\quad\quad = \dfrac{1}{2} + \dfrac{\sqrt{3}}{2}$

$\quad\quad\quad\quad\quad = \dfrac{1 + \sqrt{3}}{2}$

$1 \neq \dfrac{1 + \sqrt{3}}{2}$

73. $\dfrac{\pi}{12} = (\dfrac{\pi}{12} \cdot \dfrac{180}{\pi})°$

$\quad\quad = 15°$

Problem Set 5.2

1. $\sin 15° = \sin(45° - 30°)$

$\quad\quad\quad = \sin 45° \cos 30° - \cos 45° \sin 30°$

$\quad\quad\quad = (\dfrac{\sqrt{2}}{2})(\dfrac{\sqrt{3}}{2}) - (\dfrac{\sqrt{2}}{2})(\dfrac{1}{2})$

$\quad\quad\quad = \dfrac{\sqrt{6}}{4} - \dfrac{\sqrt{2}}{4}$

$\quad\quad\quad = \dfrac{\sqrt{6} - \sqrt{2}}{4}$

5. $\sin \dfrac{7\pi}{12} = \sin(\dfrac{3\pi}{12} + \dfrac{4\pi}{12})$

$\quad\quad\quad = \sin(\dfrac{\pi}{4} + \dfrac{\pi}{3})$

$\quad\quad\quad = \sin \dfrac{\pi}{4} \cos \dfrac{\pi}{3} + \cos \dfrac{\pi}{4} \sin \dfrac{\pi}{3}$

$\quad\quad\quad = (\dfrac{\sqrt{2}}{2})(\dfrac{1}{2}) + (\dfrac{\sqrt{2}}{2})(\dfrac{\sqrt{3}}{2})$

$$= \frac{\sqrt{2}}{4} + \frac{\sqrt{6}}{4}$$

$$= \frac{\sqrt{2} + \sqrt{6}}{4}$$

9. $\sin(x + 2\pi) = \sin x \cos 2\pi + \cos x \sin 2\pi$
    $= \sin x(1) + \cos x(0)$
    $= \sin x$

13. $\cos(180° - \theta) = \cos 180° \cos \theta + \sin 180° \sin \theta$
    $= -1(\cos \theta) + 0(\sin \theta)$
    $= - \cos \theta$

17. $\tan(x + \frac{\pi}{4}) = \dfrac{\tan x + \tan \frac{\pi}{4}}{1 - \tan x \tan \frac{\pi}{4}}$

    $= \dfrac{\tan x + 1}{1 - \tan x(1)}$

    $= \dfrac{1 + \tan x}{1 - \tan x}$

21. $\sin 3x \cos 2x + \cos 3x \sin 2x = \sin(3x + 2x)$
    $= \sin 5x$

25. $\cos 15° \cos 75° - \sin 15° \sin 75° = \cos(15° + 75°)$
    $= \cos 90°$
    $= 0$

29. $y = 3 \cos(7x) \cos(5x) + 3 \sin(7x) \sin(5x)$
    $= 3[\cos(7x) \cos(5x) + \sin(7x) \sin(5x)]$
    $= 3 \cos(7x - 5x)$
    $y = 3 \cos 2x$

    The graph is a cosine curve with an amplitude of 3 and period $= \frac{2\pi}{2} = \pi$.

33. $y = 2(\sin x \cos \frac{\pi}{3} + \cos x \sin \frac{\pi}{3})$

    $y = 2 \sin(x + \frac{\pi}{3})$

    The graph is a sine curve with:

    Amplitude = 2

    Period = 2π

    Phase shift = $- \frac{\pi}{3}$

86

The 5 points on the x-axis we use are:

$$A = -\frac{\pi}{3}$$

$$E = -\frac{\pi}{3} + 2\pi = \frac{5\pi}{3}$$

$$C = \frac{1}{2}(-\frac{\pi}{3} + \frac{5\pi}{3}) = \frac{1}{2}(\frac{4\pi}{3}) = \frac{2\pi}{3}$$

$$B = \frac{1}{2}(-\frac{\pi}{3} + \frac{2\pi}{3}) = \frac{1}{2}(\frac{\pi}{3}) = \frac{\pi}{6}$$

$$D = \frac{1}{2}(\frac{2\pi}{3} + \frac{5\pi}{3}) = \frac{1}{2}(\frac{7\pi}{3}) = \frac{7\pi}{6}$$

$$-\frac{\pi}{3}, \frac{\pi}{6}, \frac{2\pi}{3}, \frac{7\pi}{6}, \frac{5\pi}{3}$$

The 2 points on the y-axis we use are: 2 and -2.

37. If $\sin A = \frac{1}{\sqrt{5}}$ with A in Q I, then

$$\cos A = \sqrt{1 - \sin^2 A}$$

$$= \sqrt{1 - \frac{1}{5}}$$

$$= \sqrt{\frac{4}{5}}$$

$$= \frac{2}{\sqrt{5}}$$

Also, $\tan A = \dfrac{\sin A}{\cos A}$

$$= \dfrac{\dfrac{1}{\sqrt{5}}}{\dfrac{2}{\sqrt{5}}}$$

$$= \frac{1}{2}$$

We have $\tan A = \frac{1}{2}$ and $\tan B = \frac{3}{4}$.

Therefore, $\tan(A + B) = \dfrac{\dfrac{1}{2} + \dfrac{3}{4}}{1 - \left(\dfrac{1}{2}\right)\left(\dfrac{3}{4}\right)}$

$$= \dfrac{\dfrac{5}{4}}{\dfrac{5}{8}}$$

$$= 2$$

$$\cot(A + B) = \dfrac{1}{\tan(A + B)}$$

$$= \dfrac{1}{2}$$

The angle $(A + B)$ terminates in Q I because its tangent is positive. (If its tangent were negative, it would terminate in Q II.)

41. $\sin 2x = \sin(x + x)$
$\qquad = \sin x \cos x + \cos x \sin x$
$\qquad = 2 \sin x \cos x$

45. $\cos(x - 90°) - \cos(x + 90°)$

$= [\cos x \cos 90° + \sin x \sin 90°]$ $\;- [\cos x \cos 90° - \sin x \sin 90°]$	Sum and difference formulas
$= [\cos x(0) + \sin x(1)]$ $\;- [\cos x(0) - \sin x(1)]$	Substitute exact values
$= \sin x - (-\sin x)$	Multiply
$= 2 \sin x$	Subtract

49.

$\dfrac{\sin(A - B)}{\cos A \cos B} = \dfrac{\sin A \cos B - \cos A \sin B}{\cos A \cos B}$	Difference formula
$= \dfrac{\sin A \cos B}{\cos A \cos B} - \dfrac{\cos A \sin B}{\cos A \cos B}$	Separate fractions
$= \dfrac{\sin A}{\cos A} - \dfrac{\sin B}{\cos B}$	Reduce
$= \tan A - \tan B$	Ratio identity

53. The graph is a sine curve with amplitude = 4 and period $= \dfrac{2\pi}{2} = \pi$.

57. The graph is a cosine curve with amplitude = 2 and period $= \dfrac{2\pi}{\pi} = 2$.

61. The graph is a cosine curve with amplitude $= \dfrac{1}{2}$ and period $= \dfrac{2\pi}{3}$.

Problem Set 5.3

1. If $\sin A = -\frac{3}{5}$ with A in Q III, then

$$\cos A = -\sqrt{1 - \sin^2 A}$$

$$= -\sqrt{1 - \frac{9}{25}}$$

$$= -\sqrt{\frac{16}{25}}$$

$$= -\frac{4}{5}$$

Therefore, $\sin 2A = 2 \sin A \cos A$

$$= 2\left(-\frac{3}{5}\right)\left(-\frac{4}{5}\right)$$

$$= \frac{24}{25}$$

5. If $\cos x = \frac{1}{\sqrt{10}}$ with x in Q IV, then

$$\cos 2x = 2 \cos^2 x - 1$$

$$= 2\left(\frac{1}{\sqrt{10}}\right)^2 - 1$$

$$= 2\left(\frac{1}{10}\right) - 1$$

$$= \frac{1}{5} - 1$$

$$= -\frac{4}{5}$$

9. If $\tan \theta = \frac{5}{12}$ with θ in Q I, we can draw the triangle at the right and find the hypotenuse using the Pythagorean theorem.

$$\text{hypotenuse} = \sqrt{5^2 + 12^2}$$

$$= \sqrt{25 + 144}$$

$$= \sqrt{169}$$

$$= 13$$

Then $\sin \theta = \frac{5}{13}$ and $\cos \theta = \frac{12}{13}$.

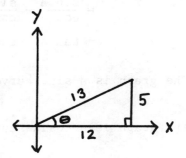

Therefore, $\sin 2\theta = 2 \sin \theta \cos \theta$

$$= 2 \left(\frac{5}{13}\right) \left(\frac{12}{13}\right)$$

$$= \frac{120}{169}$$

13. If $\csc t = \sqrt{5}$ with t in Q II, then $\sin t = \frac{1}{\sqrt{5}}$.

Therefore, $\cos 2t = 1 - 2 \sin^2 t$

$$= 1 - 2 \left[\frac{1}{\sqrt{5}}\right]^2$$

$$= 1 - \frac{2}{5}$$

$$= \frac{3}{5}$$

17. $y = 4 - 8 \sin^2 x$
$y = 4(1 - 2 \sin^2 x)$
$y = 4 \cos 2x$

The graph is a cosine curve with amplitude = 4 and period = $\frac{2\pi}{2} = \pi$.

21. $y = 1 - 2 \sin^2 2x$
$y = \cos 2(2x)$
$y = \cos 4x$

The graph is a cosine curve with amplitude = 1 and period = $\frac{2\pi}{4} = \frac{\pi}{2}$.

25. $\cos 120° = -\cos 60°$

$$= -\frac{1}{2}$$

$\cos^2 60° - \sin^2 60° = \left(\frac{1}{2}\right)^2 - \left(\frac{\sqrt{3}}{2}\right)^2$

$$= \frac{1}{4} - \frac{3}{4}$$

$$= -\frac{1}{2}$$

Therefore, they are equal.

29. $2 \sin 15° \cos 15° = \sin 2(15°)$

$$= \sin 30°$$

$$= \frac{1}{2}$$

33. $\sin \dfrac{\pi}{12} \cos \dfrac{\pi}{12} = \dfrac{1}{2} \left(2 \sin \dfrac{\pi}{12} \cos \dfrac{\pi}{12}\right)$

$\qquad\qquad\qquad = \dfrac{1}{2} \sin 2\left(\dfrac{\pi}{12}\right)$

$\qquad\qquad\qquad = \dfrac{1}{2} \sin \dfrac{\pi}{6}$

$\qquad\qquad\qquad = \dfrac{1}{2} \left(\dfrac{1}{2}\right)$

$\qquad\qquad\qquad = \dfrac{1}{4}$

37. $(\sin x - \cos x)^2 = \sin^2 x - 2 \sin x \cos x + \cos^2 x$ Expand
$\qquad\qquad\qquad = (\sin^2 x + \cos^2 x) - 2 \sin x \cos x$ Commutative property
$\qquad\qquad\qquad = 1 - 2 \sin x \cos x$ Pythagorean identity
$\qquad\qquad\qquad = 1 - \sin 2x$ Double-angle identity

41. $\dfrac{\sin 2\theta}{1 - \cos 2\theta} = \dfrac{2 \sin \theta \cos \theta}{1 - (1 - 2 \sin^2 \theta)}$ Double-angle identities

$\qquad\qquad\quad = \dfrac{2 \sin \theta \cos \theta}{2 \sin^2 \theta}$ Subtract

$\qquad\qquad\quad = \dfrac{\cos \theta}{\sin \theta}$ Reduce

$\qquad\qquad\quad = \cot \theta$ Ratio identity

45. $\sin 3\theta = \sin(2\theta + \theta)$ Addition
$\qquad\quad = \sin 2\theta \cos \theta + \cos 2\theta \sin \theta$ Sum formula
$\qquad\quad = 2 \sin \theta \cos \theta \cos \theta + (1 - 2 \sin^2 \theta) \sin \theta$ Double-angle identities
$\qquad\quad = 2 \sin \theta \cos^2 \theta + \sin \theta - 2 \sin^3 \theta$ Multiply
$\qquad\quad = 2 \sin \theta(1 - \sin^2 \theta) + \sin \theta - 2 \sin^3 \theta$ Pythagorean identity
$\qquad\quad = 2 \sin \theta - 2 \sin^3 \theta + \sin \theta - 2 \sin^3 \theta$ Multiply
$\qquad\quad = 3 \sin \theta - 4 \sin^3 \theta$ Add like terms

49. $\dfrac{\cos 2\theta}{\sin \theta \cos \theta} = \dfrac{\cos^2 \theta - \sin^2 \theta}{\sin \theta \cos \theta}$ Double-angle identity

$\qquad\qquad\quad = \dfrac{\cos^2 \theta}{\sin \theta \cos \theta} - \dfrac{\sin^2 \theta}{\sin \theta \cos \theta}$ Separate into 2 fractions

$\qquad\qquad\quad = \dfrac{\cos \theta}{\sin \theta} - \dfrac{\sin \theta}{\cos \theta}$ Reduce

$\qquad\qquad\quad = \cot \theta - \tan \theta$ Ratio identity

53. $\dfrac{\cos B - \sin B \tan B}{\sec B \sin 2B} = \dfrac{\cos B - \sin B \left(\dfrac{\sin B}{\cos B}\right)}{\dfrac{1}{\cos B}\sin 2B}$ Ratio identity and reciprocal identity

$= \dfrac{\cos B\left(\cos B - \dfrac{\sin^2 B}{\cos B}\right)}{\cos B\left(\dfrac{\sin 2B}{\cos B}\right)}$ Multiply numerator and denominator by LCD

$= \dfrac{\cos^2 B - \sin^2 B}{\sin 2B}$ Multiply

$= \dfrac{\cos 2B}{\sin 2B}$ Double-angle identity

$= \cot 2B$ Ratio identity

57. $\dfrac{1 - \tan x}{1 + \tan x} = \dfrac{1 - \dfrac{\sin x}{\cos x}}{1 + \dfrac{\sin x}{\cos x}}$ Ratio identity

$= \dfrac{\cos x\left(1 - \dfrac{\sin x}{\cos x}\right)}{\cos x\left(1 + \dfrac{\sin x}{\cos x}\right)}$ Multiply numerator and denominator by LCD

$= \dfrac{\cos x - \sin x}{\cos x + \sin x}$ Distributive property

$= \left(\dfrac{\cos x - \sin x}{\cos x + \sin x}\right) \cdot \left(\dfrac{\cos x - \sin x}{\cos x - \sin x}\right)$ Multiply by a fraction equal to 1

$= \dfrac{\cos^2 x - 2\sin x \cos x + \sin^2 x}{\cos^2 x - \sin^2 x}$ Multiply fractions

$= \dfrac{1 - 2\sin x \cos x}{\cos^2 x - \sin^2 x}$ Pythagorean identity

$= \dfrac{1 - \sin 2x}{\cos 2x}$ Double-angle identities

61. Let $y_1 = 3$ (a horizontal line) and $y_2 = -3\cos x$ (a cosine curve reflected about the x-axis with amplitude of 3). Graph y_1, y_2, and $y = y_1 + y_2$ on the same coordinate system.

65. Let $y_1 = \dfrac{1}{2}x$ (a line) and $y_2 = \sin \pi x$ (a sine curve with amplitude = 1 and period $= \dfrac{2\pi}{\pi} = 2$). Graph y_1, y_2, and $y = y_1 + y_2$ on the same coordinate system.

92

1. If A is in Q IV, then $270° < A < 360°$ and $135° < \frac{A}{2} < 180°$. Therefore, $\frac{A}{2}$ is in Q II.

$$\sin \frac{A}{2} = \sqrt{\frac{1 - \cos A}{2}}$$

$$= \sqrt{\frac{1 - 1/2}{2}}$$

$$= \sqrt{\frac{1}{4}}$$

$$= \frac{1}{2}$$

5. If A is in Q III, then $180° < A < 270°$ and $90° < \frac{A}{2} < 135°$. Therefore, $\frac{A}{2}$ is in Q II.

If $\sin A = -\frac{3}{5}$, then $\cos A = -\sqrt{1 - \sin^2 A}$

$$= -\sqrt{1 - \frac{9}{25}}$$

$$= -\sqrt{\frac{16}{25}}$$

$$= -\frac{4}{5}$$

Therefore, $\cos \frac{A}{2} = -\sqrt{\frac{1 + \cos A}{2}}$

$$= -\sqrt{\frac{1 - 4/5}{2}}$$

$$= -\sqrt{\frac{1/5}{2}}$$

$$= -\sqrt{\frac{1}{10}}$$

$$= -\frac{1}{\sqrt{10}}$$

9. If B is in Q III, $\frac{B}{2}$ is in Q II. (See problem 5 above.)

If $\sin B = -\frac{1}{3}$, then $\cos B = -\sqrt{1 - \sin^2 B}$

$$= -\sqrt{1 - \frac{1}{9}}$$

$$= -\sqrt{\frac{8}{9}}$$

$$= -\frac{2\sqrt{2}}{3}$$

Therefore, $\sin \dfrac{B}{2} = \sqrt{\dfrac{1 - \cos B}{2}}$

$$= \sqrt{\dfrac{1 + \dfrac{2\sqrt{2}}{3}}{2}}$$ (Multiply numerator and denominator by 3)

$$= \sqrt{\dfrac{3 + 2\sqrt{2}}{6}}$$

13. Using the information from problem 9 above:

$$\tan \dfrac{B}{2} = \dfrac{1 - \cos B}{\sin B}$$

$$= \dfrac{1 + \dfrac{2\sqrt{2}}{3}}{-\dfrac{1}{3}}$$

$$= \dfrac{3 + 2\sqrt{2}}{-1}$$

$$= -3 - 2\sqrt{2}$$

17. $\cos 2A = 1 - 2 \sin^2 A$

$$= 1 - 2\left(\dfrac{4}{5}\right)^2$$

$$= 1 - \dfrac{32}{25}$$

$$= -\dfrac{7}{25}$$

21. If B is in Q I, then $\dfrac{B}{2}$ must be in Q I.

If $\sin B = \dfrac{3}{5}$, then $\cos B = \sqrt{1 - \sin^2 B}$

$$= \sqrt{1 - \dfrac{9}{25}} = \sqrt{\dfrac{16}{25}} = \dfrac{4}{5}$$

Therefore, $\cos \dfrac{B}{2} = \sqrt{\dfrac{1 + \cos B}{2}}$

$$= \sqrt{\dfrac{1 + \dfrac{4}{5}}{2}} = \sqrt{\dfrac{9}{10}} = \dfrac{3}{\sqrt{10}}$$

25. If $\sin A = \frac{4}{5}$, then $\cos A = -\sqrt{1 - \sin^2 A}$

$$= -\sqrt{1 - \frac{16}{25}} = -\sqrt{\frac{9}{25}} = -\frac{3}{5}$$

We have $\qquad \sin A = \frac{4}{5} \qquad \sin B = \frac{3}{5}$

$\qquad\qquad \cos A = -\frac{3}{5} \qquad \cos B = \frac{4}{5}$ (From #21)

Therefore, $\cos (A - B) = \cos A \cos B + \sin A \sin B$

$$= -\frac{3}{5}\left(\frac{4}{5}\right) + \frac{4}{5}\left(\frac{3}{5}\right)$$

$$= -\frac{12}{25} + \frac{12}{25}$$

$$= 0$$

29. $y = 2 \cos^2 \left(\frac{x}{2}\right)$

$$y = 2\left(\pm\sqrt{\frac{1 + \cos x}{2}}\right)^2$$

$$y = 2\left(\frac{1 + \cos x}{2}\right)$$

$$y = 1 + \cos x$$

Let $y_1 = 1$ (a horizontal line) and $y_2 = \cos x$. Then graph y_1, y_2, and $y = y_1 + y_2$ on the same coordinate system.

33. $\sin 75° = \sin \left(\frac{150°}{2}\right)$

$$= \sqrt{\frac{1 - \cos 150°}{2}}$$

$$= \sqrt{\frac{1 - \left(-\frac{\sqrt{3}}{2}\right)}{2}}$$

$$= \sqrt{\frac{2 + \sqrt{3}}{4}}$$

$$= \frac{\sqrt{2 + \sqrt{3}}}{2}$$

37. $\dfrac{\csc \theta - \cot \theta}{2 \csc \theta} = \dfrac{\dfrac{1}{\sin \theta} - \dfrac{\cos \theta}{\sin \theta}}{\dfrac{2}{\sin \theta}}$ Reciprocal and ratio identities

$= \dfrac{\dfrac{1 - \cos \theta}{\sin \theta}}{\dfrac{2}{\sin \theta}}$ Subtract

$= \dfrac{1 - \cos \theta}{2}$ Divide

$= \sin^2 \dfrac{\theta}{2}$ Half-angle formula

41. $\tan \dfrac{B}{2} = \dfrac{1 - \cos B}{\sin B}$ Half-angle formula

$= \dfrac{1}{\sin B} - \dfrac{\cos B}{\sin B}$ Separate fractions

$= \csc B - \cot B$ Reciprocal and ratio identities

45. $\dfrac{\tan \theta + \sin \theta}{2 \tan \theta} = \dfrac{\dfrac{\sin \theta}{\cos \theta} + \sin \theta}{\dfrac{2 \sin \theta}{\cos \theta}}$ Ratio identity

$= \dfrac{\sin \theta + \sin \theta \cos \theta}{2 \sin \theta}$ Multiply numerator and denominator by $\cos \theta$

$= \dfrac{\sin \theta (1 + \cos \theta)}{2 \sin \theta}$ Factor

$= \dfrac{1 + \cos \theta}{2}$ Reduce

$= \cos^2 \dfrac{\theta}{2}$ Half-angle formula

49. Let $\theta = \text{Arcsin } \dfrac{3}{5}$. Then $\sin \theta = \dfrac{3}{5}$ and $-90° \leqslant \theta \leqslant 90°$.

Next we draw a triangle and label the opposite side and hypotenuse.

Using the figure to the right, $\sin \theta = \dfrac{3}{5}$.

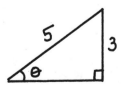

53. Let $\theta = \text{Tan}^{-1} x$. Then $\tan \theta = \dfrac{x}{1}$ and $-90° \leqslant \theta \leqslant 90°$.

Next we draw a triangle and label the opposite and adjacent sides. Then we find the hypotenuse using the Pythagorean theorem.

hypotenuse $= \sqrt{x^2 + 1^2} = \sqrt{x^2 + 1}$

Using the figure to the right, $\sin \theta = \dfrac{x}{\sqrt{x^2 + 1}}$

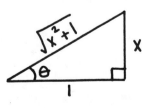

<u>Problem Set 5.5</u>

1. Let $\alpha = \text{Arcsin } \frac{3}{5}$ and $\beta = \text{Arctan } 2$.

    Then $\sin \alpha = \frac{3}{5}$ and $-90° \leqslant \alpha \leqslant 90°$ and $\tan \beta = 2$ and $-90° \leqslant \beta \leqslant 90°$

    Also, $\sin (\text{Arcsin } \frac{3}{5} - \text{Arctan } 2) = \sin(\alpha - \beta)$
    $$= \sin \alpha \cos \beta - \cos \alpha \sin \beta$$

    Drawing and labeling a triangle for α and another for β we have

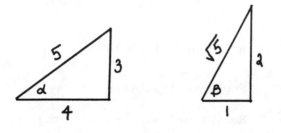

    Using the Pythagorean theorem we find the missing sides.

    From the figure above we have

    $$\sin \alpha = \frac{3}{5} \qquad \sin \beta = \frac{2}{\sqrt{5}}$$

    $$\cos \alpha = \frac{4}{5} \qquad \cos \beta = \frac{1}{\sqrt{5}}$$

    Substituting these above, we get

    $$\sin \alpha \cos \beta - \cos \alpha \sin \beta = \frac{3}{5}\left(\frac{1}{\sqrt{5}}\right) - \frac{4}{5}\left(\frac{2}{\sqrt{5}}\right)$$

    $$= \frac{3}{5\sqrt{5}} - \frac{8}{5\sqrt{5}}$$

    $$= -\frac{5}{5\sqrt{5}}$$

    $$= -\frac{1}{\sqrt{5}}$$

5. Let $\alpha = \text{Cos}^{-1} \dfrac{1}{\sqrt{5}}$. Then $\cos \alpha = \dfrac{1}{\sqrt{5}}$ and $0° \leqslant \alpha \leqslant 180°$.

Also, $\sin\left(2 \, \text{Cos}^{-1} \dfrac{1}{\sqrt{5}}\right) = \sin 2\alpha$

$$= 2 \sin \alpha \cos \alpha$$

Drawing and labeling a triangle for α, we have:

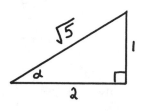

Using the Pythagorean theorem, we find the adjacent side.

adjacent side $= \sqrt{(\sqrt{5})^2 - 1^2} = \sqrt{4} = 2$

From the figure above, we have

$$\sin \alpha = \frac{1}{\sqrt{5}}$$

$$\cos \alpha = \frac{2}{\sqrt{5}}$$

Substituting these above, we get

$$2 \sin \alpha \cos \alpha = 2\left(\frac{1}{\sqrt{5}}\right)\left(\frac{2}{\sqrt{5}}\right)$$

$$= \frac{4}{5}$$

9. Let $\alpha = \text{Sin}^{-1} x$. Then $\sin \alpha = \dfrac{x}{1}$ and $-90° \leqslant \alpha \leqslant 90°$.

Also, $\sin(2 \, \text{Sin}^{-1} x) = \sin 2\alpha$
$$= 2 \sin \alpha \cos \alpha$$

Drawing and labeling a triangle for α, we have:

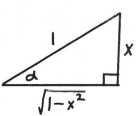

Using the Pythagorean theorem, we find the adjacent side:

adjacent side $= \sqrt{1^2 - x^2} = \sqrt{1 - x^2}$

From the figure above, we have

$$\sin \alpha = x$$
$$\cos \alpha = \sqrt{1 - x^2}$$

Substituting these above, we get

$$2 \sin \alpha \cos \alpha = 2x\sqrt{1 - x^2}$$

13. $\sin 30° \sin 120° = \frac{1}{2}[\cos(-90°) - \cos 150°]$

$\frac{1}{2} \cdot \frac{\sqrt{3}}{2} = \frac{1}{2}[0 - (-\frac{\sqrt{3}}{2})]$

$\frac{\sqrt{3}}{4} = \frac{\sqrt{3}}{4}$

17. $\cos 8x \cos 2x = \frac{1}{2}[\cos(8x + 2x) + \cos(8x - 2x)]$

$= \frac{1}{2}(\cos 10x + \cos 6x)$

21. $\sin 4\pi \sin 2\pi = \frac{1}{2}[\cos(4\pi - 2\pi) - \cos(4\pi + 2\pi)]$

$= \frac{1}{2}[\cos 2\pi - \cos 6\pi]$

$= \frac{1}{2}[1 - 1]$

$= 0$

25. $\sin 7x + \sin 3x = 2 \sin \frac{7x + 3x}{2} \cos \frac{7x - 3x}{2}$

$= 2 \sin 5x \cos 2x$

29. $\sin \frac{7\pi}{12} - \sin \frac{\pi}{12} = 2 \cos \frac{7\pi/12 + \pi/12}{2} \sin \frac{7\pi/12 - \pi/12}{2}$

$= 2 \cos \frac{\pi}{3} \sin \frac{\pi}{4}$

$= 2(\frac{1}{2})\left[\frac{1}{\sqrt{2}}\right]$

$= \frac{1}{\sqrt{2}}$

33. $\dfrac{\sin 4x + \sin 6x}{\cos 4x - \cos 6x} = \dfrac{2 \sin 5x \cos(-x)}{-2 \sin 5x \sin(-x)}$ Sum to product formulas

$= -\dfrac{\cos(-x)}{\sin(-x)}$ Reduce

$= -\dfrac{\cos x}{-\sin x}$ Cosine is even function, sine is odd function

$= \cot x$ Ratio identity

37. The graph is a sine curve with a phase shift $= -\frac{\pi}{4}$.

41. Amplitude = 1

Period = $\frac{2\pi}{2}$ = π

Phase shift = $\frac{\pi/2}{2}$ = $\frac{\pi}{4}$

The 5 points we use on the x-axis are:

A = $\frac{\pi}{4}$

E = $\frac{\pi}{4} + \pi$ = $\frac{5\pi}{4}$

C = $\frac{1}{2}(\frac{\pi}{4} + \frac{5\pi}{4})$ = $\frac{3\pi}{4}$

B = $\frac{1}{2}(\frac{\pi}{4} + \frac{3\pi}{4})$ = $\frac{\pi}{2}$

D = $\frac{1}{2}(\frac{3\pi}{4} + \frac{5\pi}{4})$ = π

The 2 points we use on the y-axis are:

1 and -1.

45. Amplitude = 3

Period = $\frac{2\pi}{\pi}$ = 2

Phase shift = $\frac{\pi/2}{\pi}$ = $\frac{1}{2}$

The 5 points we use on the x-axis are:

A = $\frac{1}{2}$

E = $\frac{1}{2} + 2$ = $\frac{5}{2}$

C = $\frac{1}{2}(\frac{1}{2} + \frac{5}{2})$ = $\frac{3}{2}$

B = $\frac{1}{2}(\frac{1}{2} + \frac{3}{2})$ = 1

D = $\frac{1}{2}(\frac{3}{2} + \frac{5}{2})$ = 2

The 2 points we use on the y-axis are:

3 and -3.

Chapter 5 Test

1. $\sin \theta \sec \theta = \sin \theta \cdot \dfrac{1}{\cos \theta}$ Reciprocal identity

      $= \dfrac{\sin \theta}{\cos \theta}$ Multiply

      $= \tan \theta$ Ratio identity

5. $\dfrac{\cos t}{1 - \sin t} = \dfrac{\cos t}{1 - \sin t} \cdot \dfrac{1 + \sin t}{1 + \sin t}$ Multiply by a fraction equal to 1

      $= \dfrac{\cos t(1 + \sin t)}{1 - \sin^2 t}$ Multiply

      $= \dfrac{\cos t(1 + \sin t)}{\cos^2 t}$ Pythagorean identity

      $= \dfrac{1 + \sin t}{\cos t}$ Reduce

9. $\cos^4 A - \sin^4 A = (\cos^2 A + \sin^2 A)(\cos^2 A - \sin^2 A)$ Factor

      $= 1(\cos^2 A - \sin^2 A)$ Pythagorean identity

      $= \cos 2A$ Double-angle formula

13. A is in Q IV. Then $\cos A = \sqrt{1 - \sin^2 A}$

      $= \sqrt{1 - \dfrac{9}{25}} = \sqrt{\dfrac{16}{25}} = \dfrac{4}{5}$

B is in Q II. Then $\cos B = -\sqrt{1 - \sin^2 A}$

      $= -\sqrt{1 - \dfrac{144}{169}} = -\sqrt{\dfrac{25}{169}} = -\dfrac{5}{13}$

We have $\sin A = -\dfrac{3}{5}$ $\sin B = \dfrac{12}{13}$

            $\cos A = \dfrac{4}{5}$ $\cos B = -\dfrac{5}{13}$

Therefore, $\sin (A + B) = \sin A \cos B + \cos A \sin B$

      $= -\dfrac{3}{5}\left(-\dfrac{5}{13}\right) + \dfrac{4}{5}\left(\dfrac{12}{13}\right)$

      $= \dfrac{15}{65} + \dfrac{48}{65}$

      $= \dfrac{63}{65}$

17. Using the information from problem 13 above:

If $270° \leqslant A \leqslant 360°$, then $135° \leqslant \frac{A}{2} \leqslant 180°$. Therefore, $\frac{A}{2}$ is in Q II.

$$\sin \frac{A}{2} = \sqrt{\frac{1 - \cos A}{2}}$$

$$= \sqrt{\frac{1 - \frac{4}{5}}{2}}$$

$$= \sqrt{\frac{1}{10}}$$

$$= \frac{1}{\sqrt{10}}$$

21. $\tan \frac{\pi}{12} = \tan \left(\frac{\pi}{3} - \frac{\pi}{4}\right)$

$$= \frac{\tan \frac{\pi}{3} - \tan \frac{\pi}{4}}{1 + \tan \frac{\pi}{3} \tan \frac{\pi}{4}}$$

$$= \frac{\sqrt{3} - 1}{1 + (\sqrt{3})(1)}$$

$$= \frac{\sqrt{3} - 1}{\sqrt{3} + 1}$$

25. $\cos 2A = 1 - 2 \sin^2 A$

$$= 1 - 2\left(-\frac{1}{\sqrt{5}}\right)^2$$

$$= 1 - \frac{2}{5}$$

$$= \frac{3}{5}$$

For $\cos \frac{A}{2}$, we must find $\cos A$.

$\cos A = -\sqrt{1 - \sin^2 A}$

$$= -\sqrt{1 - \frac{1}{5}} = -\sqrt{\frac{4}{5}} = -\frac{2}{\sqrt{5}} = -\frac{2\sqrt{5}}{5}$$

If $180° \leqslant A \leqslant 270°$, then $90° \leqslant \frac{A}{2} \leqslant 135°$.

$$\cos \frac{A}{2} = -\sqrt{\frac{1 + \cos A}{2}}$$

$$= -\sqrt{\frac{1 + (-\frac{2\sqrt{5}}{5})}{2}}$$

$$= -\sqrt{\frac{5 - 2\sqrt{5}}{10}}$$

29. Let $\alpha = \text{Arcsin } \frac{4}{5}$ and $\beta = \text{Arctan } 2$.

Then we can draw 2 triangles and label α and β accordingly:

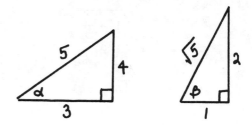

From the figures to the right, we have

$$\sin \alpha = \frac{4}{5} \qquad \sin \beta = \frac{2}{\sqrt{5}}$$

$$\cos \alpha = \frac{3}{5} \qquad \cos \beta = \frac{1}{\sqrt{5}}$$

Therefore, $\cos(\alpha - \beta) = \cos \alpha \cos \beta + \sin \alpha \sin \beta$

$$= \frac{3}{5}\left(\frac{1}{\sqrt{5}}\right) + \frac{4}{5}\left(\frac{2}{\sqrt{5}}\right)$$

$$= \frac{3}{5\sqrt{5}} + \frac{8}{5\sqrt{5}}$$

$$= \frac{11}{5\sqrt{5}}$$

33. $\sin 6x \sin 4x = \frac{1}{2}[\cos(6x - 4x) - \cos(6x + 4x)]$

$$= \frac{1}{2}(\cos 2x - \cos 10x)$$

EQUATIONS

Problem Set 6.1

1. $2 \sin \theta = 1$

   $\sin \theta = \dfrac{1}{2}$ Divide both sides by 2

   $\theta = 30°$ or $150°$ $\hat{\theta} = 30°$ and θ is in Q I or Q II

5. $2 \tan \theta + 2 = 0$
      $2 \tan \theta = -2$ Subtract 2 from both sides
        $\tan \theta = -1$ Divide both sides by 2
           $\theta = 135°$ or $315°$ $\hat{\theta} = 45°$ and θ is in Q II or Q IV

9. $2 \cos t = 6 \cos t - \sqrt{12}$

   $-4 \cos t = -2\sqrt{3}$ Subtract $6 \cos t$ from both sides

   $\cos t = \dfrac{\sqrt{3}}{2}$ Divide both sides by -4

   $t = \dfrac{\pi}{6}$ or $\dfrac{11\pi}{6}$ $\hat{t} = \dfrac{\pi}{6}$ and t is in Q I or Q IV

13. $4 \sin \theta - 3 = 0$
        $4 \sin \theta = 3$ Add 3 to both sides
          $\sin \theta = 0.75$ Divide both sides by 4
            $\theta = 48.6°$ or $131.4°$ $\hat{\theta} = 48.6°$ and θ is in Q I or Q II

17. $\sin \theta - 3 = 5 \sin \theta$
         $-3 = 4 \sin \theta$ Subtract $\sin \theta$ from both sides
       $\sin \theta = -0.75$ Divide both sides by 4
         $\theta = 228.6°$ or $311.4°$ $\hat{\theta} = 48.6°$ and θ is in Q III or Q IV

21. $\tan x(\tan x - 1) = 0$

    $\tan x = 0$ or $\tan x - 1 = 0$ Set each factor = 0

    $x = 0$ or π $\tan x = 1$ Solve each resulting equation

                    $x = \dfrac{\pi}{4}$ or $\dfrac{5\pi}{4}$

25. $2 \sin^2 x - \sin x - 1 = 0$ Standard form
      $(2 \sin x + 1)(\sin x - 1) = 0$ Factor
      $2 \sin x + 1 = 0$ or $\sin x - 1 = 0$ Set each factor = 0
      $2 \sin x = -1$ $\sin x = 1$ Solve each resulting equation

        $\sin x = -\dfrac{1}{2}$ $x = \dfrac{\pi}{2}$

         $x = \dfrac{7\pi}{6}$ or $\dfrac{11\pi}{6}$

29. $\sqrt{3} \tan \theta - 2 \sin \theta \tan \theta = 0$ Standard form
    $\tan \theta (\sqrt{3} - 2 \sin \theta) = 0$ Factor
    $\tan \theta = 0$ or $\sqrt{3} - 2 \sin \theta = 0$ Set each factor = 0
       $\theta = 0°$ or $180°$ $-2 \sin \theta = -\sqrt{3}$ Solve each resulting equation

$$\sin \theta = \frac{\sqrt{3}}{2}$$

$$\theta = 60° \text{ or } 120°$$

33. $2 \sin^2 \theta - 2 \sin \theta - 1 = 0$ where $a = 2$, $b = -2$, $c = -1$

$$\sin \theta = \frac{-(-2) \pm \sqrt{(-2)^2 - 4(2)(-1)}}{2(2)}$$

$$= \frac{2 \pm \sqrt{12}}{4}$$

$$= \frac{2 \pm 2\sqrt{3}}{4}$$

$$= \frac{1 \pm \sqrt{3}}{2}$$

$$= \frac{1 \pm 1.7321}{2}$$

$\sin \theta = 1.3666$ or $\sin \theta = -0.3660$
  No solution $\hat{\theta} = 21.5°$ and θ is in Q III or Q IV
                    $\theta = 201.5°$ or $338.5°$

37. $2 \sin^2 \theta + 1 = 4 \sin \theta$
    $2 \sin^2 \theta - 4 \sin \theta + 1 = 0$
    $a = 2$, $b = -4$, $c = 1$

$$\sin \theta = \frac{-(-4) \pm \sqrt{(-4)^2 - 4(2)(1)}}{2(2)}$$

$$= \frac{4 \pm \sqrt{8}}{4}$$

$$= \frac{4 \pm 2\sqrt{2}}{4}$$

$$= \frac{2 \pm \sqrt{2}}{2}$$

$$= \frac{2 \pm 1.4142}{2}$$

$\sin \theta = 1.7071$ or $\sin \theta = 0.2929$
  No solution $\theta = 17.0°$ or $163.0°$

41. $4 \sin t - \sqrt{3} = 2 \sin t$

$\qquad -\sqrt{3} = -2 \sin t \qquad$ Subtract 4 sin t from both sides

$\qquad \sin t = \dfrac{\sqrt{3}}{2} \qquad$ Divide both sides by -2

$\qquad t = \dfrac{\pi}{3} + 2k\pi \qquad \hat{t} = \dfrac{\pi}{3}$ and t is in Q I or Q II

$\qquad$ or $\dfrac{2\pi}{3} + 2k\pi$

45. In problem 13 we found that $\theta = 48.6°$ or $131.4°$.
Therefore, $\theta = 48.6° + 360°k$ or
$\qquad\qquad 131.4° + 360°k$

49. $\sin(3A + 30°) = \dfrac{1}{2}$

$3A + 30° = 30° + 360°k$ or $3A + 30° = 150° + 360°k$
$\qquad 3A = 360°k \qquad\qquad\qquad 3A = 120° + 360°k$
$\qquad A = 120°k \qquad\qquad\qquad\ A = 40° + 120°k$

53. $\sin(5A + 15°) = -\dfrac{1}{\sqrt{2}}$

$5A + 15° = 225° + 360°k$ or $5A + 15° = 315° + 360°k$
$\qquad 5A = 210° + 360°k \qquad\qquad 5A = 300° + 360°k$
$\qquad A = 42° + 72°k \qquad\qquad\quad A = 60° + 72°k$

57. $h = -16t^2 + vt \sin \theta$
$\quad = -16(2)^2 + (1500)(2) \sin 30°$

$\quad = -64 + 3000(\dfrac{1}{2})$

$\quad = -64 + 1500$
$\quad = 1436$ ft

65. $\sin(\theta + 45°) = \sin \theta \cos 45° + \cos \theta \sin 45°$

$\qquad\qquad = \sin \theta \left[\dfrac{1}{\sqrt{2}}\right] + \cos \theta \left[\dfrac{1}{\sqrt{2}}\right]$

$\qquad\qquad = \dfrac{1}{\sqrt{2}} \sin \theta + \dfrac{1}{\sqrt{2}} \cos \theta$

69. $\dfrac{1 - \tan^2 x}{1 + \tan^2 x} = \dfrac{1 - \dfrac{\sin^2 x}{\cos^2 x}}{1 + \dfrac{\sin^2 x}{\cos^2 x}} \qquad$ Ratio identity

$\qquad = \dfrac{\cos^2 x - \sin^2 x}{\cos^2 x + \sin^2 x} \qquad$ Multiply numerator and denominator by $\cos^2 x$

$\qquad = \dfrac{\cos 2x}{1} \qquad$ Double-angle formula and Pythagorean identity

$\qquad = \cos 2x \qquad$ Simplify

Problem Set 6.2

1. $\sqrt{3}\ \sec\ \theta = 2$

$\sec\ \theta = \dfrac{2}{\sqrt{3}}$ Divide both sides by $\sqrt{3}$

$\cos\ \theta = \dfrac{\sqrt{3}}{2}$ $\cos\ \theta = \dfrac{1}{\sec\ \theta}$

$\theta = 30°$ or $330°$ $\hat{\theta} = 30°$ and θ is in Q I or Q IV

5. $4\ \sin\ \theta - 2\ \csc\ \theta = 0$

$4\ \sin\ \theta - \dfrac{2}{\sin\ \theta} = 0$ $\csc\ \theta = \dfrac{1}{\sin\ \theta}$

$4\ \sin^2\ \theta - 2 = 0$ Multiply both sides by $\sin\ \theta$

$4\ \sin^2\ \theta = 2$ Add 2 to both sides

$\sin^2\ \theta = \dfrac{1}{2}$ Divide both sides by 4

$\sin\ \theta = \pm\dfrac{1}{\sqrt{2}}$ Take square root of both sides

$\theta = 45°,\ 135°,\ 225°,\ 315°$ $\hat{\theta} = 45°$

9. $\sin\ 2\theta - \cos\ \theta = 0$

   $2\ \sin\ \theta\ \cos\ \theta - \cos\ \theta = 0$ $\sin\ 2\theta = 2\ \sin\ \theta\ \cos\ \theta$

     $\cos\ \theta(2\ \sin\ \theta - 1) = 0$ Factor out $\cos\ \theta$

$\cos\ \theta = 0$ or $2\ \sin\ \theta - 1 = 0$ Set each factor = 0

   $\theta = 90°,\ 270°$ $2\ \sin\ \theta = 1$ Solve each equation

                    $\sin\ \theta = \dfrac{1}{2}$

               $\theta = 30°$ or $150°$

13. $\cos\ 2x - 3\ \sin\ x - 2 = 0$

   $1 - 2\ \sin^2\ x - 3\ \sin\ x - 2 = 0$ $\cos\ 2x = 1 - 2\ \sin^2\ x$

     $2\ \sin^2\ x + 3\ \sin\ x + 1 = 0$ Multiply both sides by -1

   $(2\ \sin\ x + 1)(\sin\ x + 1) = 0$ Factor

$2\ \sin\ x + 1 = 0$ or $\sin\ x + 1 = 0$ Set each factor = 0

    $2\ \sin\ x = -1$ $\sin\ x = -1$ Solve each equation

     $\sin\ x = -\dfrac{1}{2}$ $x = \dfrac{3\pi}{2}$

      $x = \dfrac{7\pi}{6}$ or $\dfrac{11\pi}{6}$

17.
$$2 \cos^2 x + \sin x - 1 = 0$$
$$2(1 - \sin^2 x) + \sin x - 1 = 0 \qquad \cos^2 x = 1 - \sin^2 x$$
$$2 - 2 \sin^2 x + \sin x - 1 = 0 \qquad \text{Simplify}$$
$$2 \sin^2 x - \sin x - 1 = 0 \qquad \text{Multiply both sides by } -1$$
$$(2 \sin x + 1)(\sin x - 1) = 0 \qquad \text{Factor}$$

$$2 \sin x + 1 = 0 \qquad \text{or } \sin x - 1 = 0 \qquad \text{Set each factor} = 0$$
$$2 \sin x = -1 \qquad \qquad \sin x = 1 \qquad \text{Solve each equation}$$

$$\sin x = -\frac{1}{2} \qquad \qquad x = \frac{\pi}{2}$$

$$x = \frac{7\pi}{6} \text{ or } \frac{11\pi}{6}$$

21.
$$2 \sin x + \cot x - \csc x = 0$$

$$2 \sin x + \frac{\cos x}{\sin x} - \frac{1}{\sin x} = 0 \qquad \cot x = \frac{\cos x}{\sin x} \text{ and } \csc x = \frac{1}{\sin x}$$

$$2 \sin^2 x + \cos x - 1 = 0 \qquad \text{Multiply both sides by } \sin x$$
$$2(1 - \cos^2 x) + \cos x - 1 = 0 \qquad \sin^2 x = 1 - \cos^2 x$$
$$2 - 2 \cos^2 x + \cos x - 1 = 0 \qquad \text{Simplify}$$
$$2 \cos^2 x - \cos x - 1 = 0 \qquad \text{Multiply both sides by } -1$$
$$(2 \cos x + 1)(\cos x - 1) = 0 \qquad \text{Factor}$$
$$2 \cos x + 1 = 0 \qquad \text{or } \cos x - 1 = 0 \qquad \text{Set each factor} = 0$$
$$2 \cos x = -1 \qquad \qquad \cos x = 1 \qquad \text{Solve each equation}$$
$$\cancel{x = 0}$$

$$\cos x = -\frac{1}{2} \qquad \text{Not possible because cotangent and cosecant are not}$$
$$\text{defined at } 0$$
$$x = \frac{2\pi}{3} \text{ or } \frac{4\pi}{3}$$

25.
$$\sqrt{3} \sin \theta + \cos \theta = \sqrt{3}$$

$$\cos \theta = \sqrt{3} - \sqrt{3} \sin \theta \qquad \text{Subtract } \sqrt{3} \sin \theta \text{ from both sides}$$
$$\cos \theta = \sqrt{3}(1 - \sin \theta) \qquad \text{Factor out } \sqrt{3}$$
$$\cos^2 \theta = 3(1 - 2 \sin \theta + \sin^2 \theta) \qquad \text{Square both sides}$$
$$1 - \sin^2 \theta = 3 - 6 \sin \theta + 3 \sin^2 \theta \qquad \cos^2 \theta = 1 - \sin^2 \theta$$
$$4 \sin^2 \theta - 6 \sin \theta + 2 = 0 \qquad \text{Simplify}$$
$$2 \sin^2 \theta - 3 \sin \theta + 1 = 0 \qquad \text{Divide both sides by 2}$$
$$(2 \sin \theta - 1)(\sin \theta - 1) = 0 \qquad \text{Factor}$$
$$2 \sin \theta - 1 = 0 \qquad \text{or } \sin \theta - 1 = 0 \qquad \text{Set each factor} = 0$$
$$2 \sin \theta = 1 \qquad \qquad \sin \theta = 1 \qquad \text{Solve each equation}$$

$$\sin \theta = \frac{1}{2} \qquad \qquad \theta = 90°$$

$$\theta = 30° \text{ or } \cancel{150°}$$

check: $\sqrt{3} \sin 30° + \cos 30° \overset{?}{=} \sqrt{3}$ $\qquad$ $\sqrt{3} \sin 150° + \cos 150° \overset{?}{=} \sqrt{3}$

$$\sqrt{3}(\tfrac{1}{2}) + \frac{\sqrt{3}}{2} \overset{?}{=} \sqrt{3} \qquad\qquad \sqrt{3}(\tfrac{1}{2}) + (-\frac{\sqrt{3}}{2}) \overset{?}{=} \sqrt{3}$$

$$\sqrt{3} \overset{\checkmark}{=} \sqrt{3} \qquad\qquad\qquad 0 \neq \sqrt{3}$$

$$\sqrt{3} \sin 90° + \cos 90° \overset{?}{=} \sqrt{3}$$

$$\sqrt{3}(1) + 0 \overset{?}{=} \sqrt{3}$$

$$\sqrt{3} \overset{\checkmark}{=} \sqrt{3}$$

Answers: 30° or 90°

29. $\qquad \sin \dfrac{\theta}{2} - \cos \theta = 0$

$\qquad\qquad \sin \dfrac{\theta}{2} = \cos \theta$ $\qquad$ Add cos θ to both sides

$\qquad\qquad \sin^2 \dfrac{\theta}{2} = \cos^2 \theta$ $\qquad$ Square both sides

$\qquad\qquad \dfrac{1 - \cos \theta}{2} = \cos^2 \theta$ $\qquad$ $\sin \dfrac{\theta}{2} = \pm \sqrt{\dfrac{1 - \cos \theta}{2}}$

$\qquad\qquad 1 - \cos \theta = 2 \cos \theta$ $\qquad$ Multiply both sides by 2

$\qquad 2 \cos^2 \theta + \cos \theta - 1 = 0$ $\qquad$ Subtract (1 − cos θ) from both sides

$(2 \cos \theta - 1)(\cos \theta + 1) = 0$ $\qquad$ Factor

$2 \cos \theta - 1 = 0 \qquad$ or $\cos \theta + 1 = 0$ $\qquad$ Set each factor = 0

$\qquad 2 \cos \theta = 1 \qquad\qquad\qquad \cos \theta = -1$ $\qquad$ Solve each equation

$\qquad\qquad\qquad\qquad\qquad \cancel{\theta = 180°}$

$\qquad\qquad \cos \theta = \dfrac{1}{2}$

$\qquad\qquad \theta = 60°$ or 300°

check: $\sin \dfrac{60°}{2} - \cos 60° \overset{?}{=} 0$ $\qquad$ $\sin \dfrac{300°}{2} - \cos 300° \overset{?}{=} 0$

$\qquad\qquad \sin 30° - \cos 60° \overset{?}{=} 0 \qquad\qquad \sin 150° - \cos 300° \overset{?}{=} 0$

$$\dfrac{1}{2} - \dfrac{1}{2} \overset{\checkmark}{=} 0 \qquad\qquad\qquad \dfrac{1}{2} - \dfrac{1}{2} \overset{\checkmark}{=} 0$$

$\qquad \sin \dfrac{180°}{2} - \cos 180° \overset{?}{=} 0$

$\qquad \sin 90° - \cos 180° \overset{?}{=} 0$

$$1 - (-1) \overset{?}{=} 0$$

$$2 \neq 0$$

Answers: 60° or 300°

33.
$$6 \cos \theta + 7 \tan \theta = \sec \theta$$

$$6 \cos \theta + \frac{7 \sin \theta}{\cos \theta} = \frac{1}{\cos \theta} \qquad \tan \theta = \frac{\sin \theta}{\cos \theta} \text{ and } \sec \theta = \frac{1}{\cos \theta}$$

$6 \cos^2 \theta + 7 \sin \theta = 1$	Multiply both sides by $\cos \theta$
$6(1 - \sin^2 \theta) + 7 \sin \theta = 1$	$\cos^2 \theta = 1 - \sin^2 \theta$
$6 - 6 \sin^2 \theta + 7 \sin \theta = 1$	Simplify
$-6 \sin^2 \theta + 7 \sin \theta + 5 = 0$	Subtract 1 from both sides
$6 \sin^2 \theta - 7 \sin \theta - 5 = 0$	Multiply both sides by -1
$(3 \sin \theta - 5)(2 \sin \theta + 1) = 0$	Factor
$3 \sin \theta - 5 = 0 \quad \text{or} \quad 2 \sin \theta + 1 = 0$	Set each factor = 0
$3 \sin \theta = 5 \qquad\qquad 2 \sin \theta = -1$	Solve each equation

$$\sin \theta = \frac{5}{3} \qquad\qquad \sin \theta = -\frac{1}{2}$$

$$\text{No solution} \qquad\qquad \theta = 210° \text{ or } 330°$$

37.
$7 \sin^2 \theta - 9 \cos 2\theta = 0$	
$7 \sin^2 \theta - 9(1 - 2 \sin^2 \theta) = 0$	$\cos 2\theta = 1 - 2 \sin^2 \theta$
$7 \sin^2 \theta - 9 + 18 \sin^2 \theta = 0$	Simplify left side
$25 \sin^2 \theta = 9$	Add 9 to both sides

$$\sin^2 \theta = \frac{9}{25} \qquad\qquad \text{Divide both sides by 25}$$

$$\sin \theta = \pm \frac{3}{5} \qquad\qquad \text{Take square root of both sides}$$

$$\sin \theta = \pm 0.6 \qquad\qquad \text{Divide right side}$$

$$\theta = 36.9°, 143.1°, \qquad \hat{\theta} = 36.9° \text{ and } \theta \text{ is in Q I,}$$
$$216.9°, \text{ or } 323.1° \qquad \text{Q II, Q III, or Q IV}$$

41. In problem 23 we get $x = \frac{\pi}{4}$. All solutions would be $x = \frac{\pi}{4} + 2k\pi$.

45. $r^4 \csc^2 \theta - R^4 \csc \theta \cot \theta =$

$$r^4 \cdot \frac{1}{\sin^2 \theta} - R^4 \cdot \frac{1}{\sin \theta} \cdot \frac{\cos \theta}{\sin \theta}$$

$$\frac{r^4}{\sin^2 \theta} - \frac{R^4 \cos \theta}{\sin^2 \theta}$$

$$\frac{r^4 - R^4 \cos \theta}{\sin^2 \theta}$$

This expression is zero only when the numerator is zero. Therefore,

$$r^4 - R^4 \cos \theta = 0$$
$$-R^4 \cos \theta = -r^4$$

$$\cos \theta = \frac{r^4}{R^4}$$

Therefore, when $\cos \theta = \frac{r^4}{R^4}$, then $r^4 \csc^2 \theta - R^4 \csc \theta \cot \theta = 0$

110

49. $\csc \dfrac{A}{2} = \dfrac{1}{\sin \dfrac{A}{2}}$

$$= \dfrac{1}{\sqrt{\dfrac{1 - \cos A}{2}}}$$

$$= \sqrt{\dfrac{2}{1 - \cos A}}$$

We know that $\sin A = \dfrac{2}{3}$ with A in Q I.

Therefore, $\cos A = \sqrt{1 - \sin^2 A}$

$$= \sqrt{1 - \dfrac{4}{9}} = \sqrt{\dfrac{5}{9}} = \dfrac{\sqrt{5}}{3}$$

Substituting this above, we get

$\csc \dfrac{A}{2} = \sqrt{\dfrac{2}{1 - \dfrac{\sqrt{5}}{3}}}$

$$= \sqrt{\dfrac{6}{3 - \sqrt{5}}}$$

53. $y = 4 \sin^2 \dfrac{x}{2}$

$y = 4\left(\dfrac{1 - \cos x}{2}\right)$

$y = 2 - 2 \cos x$

Let $y_1 = 2$ (a horizontal line) and $y_2 = -2 \cos x$ (a cosine curve reflected about the x-axis with an amplitude of 2.) Graph y_1, y_2, and $y = y_1 + y_2$ on the same coordinate axes.

Problem Set 6.3

1. $\sin 2\theta = \dfrac{\sqrt{3}}{2}$

$2\theta = 60° + 360°k$ or $2\theta = 120° + 360°k$
$\theta = 30° + 180°k$ $\theta = 60° + 180°k$

If we let k = 0 and 1, we get

$\theta = 30°$ $\theta = 60°$
$\theta = 30° + 180° = 210°$ $\theta = 60° + 180° = 240°$

Solutions: 30°, 60°, 210°, 240°

5. cos 3θ = −1

    3θ = 180° + 360°k

    θ = 60° + 120°k

If we let k = 0, 1, and 2, we get

    θ = 60°

    θ = 60° + 120° = 180°

    θ = 60° + 240° = 300°

Solutions: 60°, 180°, 300°

9. sec 3x = −1

    cos 3x = −1 $\cos 3x = \dfrac{1}{\sec 3x}$

    3x = π + 2kπ

    $x = \dfrac{\pi}{3} + \dfrac{2k\pi}{3}$

If we let k = 0, 1, and 2, we get

    $\theta = \dfrac{\pi}{3}$

    $\theta = \dfrac{\pi}{3} + \dfrac{2\pi}{3} = \pi$

    $\theta = \dfrac{\pi}{3} + \dfrac{4\pi}{3} = \dfrac{5\pi}{3}$

Solutions: $\dfrac{\pi}{3}$, π, $\dfrac{5\pi}{3}$

13. $\sin 2\theta = \dfrac{1}{2}$

    2θ = 30° + 360°k   or 2θ = 150° + 360°k

    θ = 15° + 180°k   θ = 75° + 180°k

17. $\sin 10\theta = \dfrac{\sqrt{3}}{2}$

    10θ = 60° + 360°k   or   10θ = 120° + 360°k

    θ = 6° + 36°k   θ = 12° + 36°k

21. $\cos 2x \cos x - \sin 2x \sin x = -\dfrac{\sqrt{3}}{2}$

    $\cos(2x + x) = -\dfrac{\sqrt{3}}{2}$ Sum formula

    $\cos 3x = -\dfrac{\sqrt{3}}{2}$

$$3x = \frac{5\pi}{6} + 2k\pi \qquad \text{or} \qquad 3x = \frac{7\pi}{6} + 2k\pi$$

$$x = \frac{5\pi}{18} + \frac{2k\pi}{3} \qquad\qquad x = \frac{7\pi}{18} + \frac{2k\pi}{3}$$

$$x = \frac{5\pi + 12k\pi}{18} \qquad\qquad x = \frac{7\pi + 12k\pi}{18}$$

If we let k = 0, 1, or 2, we get

$$x = \frac{5\pi}{18} \qquad\qquad x = \frac{7\pi}{18}$$

$$x = \frac{5\pi + 12\pi}{18} = \frac{17\pi}{18} \qquad\qquad x = \frac{7\pi + 12\pi}{18} = \frac{19\pi}{18}$$

$$x = \frac{5\pi + 24\pi}{18} = \frac{29\pi}{18} \qquad\qquad x = \frac{7\pi + 24\pi}{18} = \frac{31\pi}{18}$$

Solutions: $\dfrac{5\pi}{18}, \dfrac{7\pi}{18}, \dfrac{17\pi}{18}, \dfrac{19\pi}{18}, \dfrac{29\pi}{18}, \dfrac{31\pi}{18}$

25. $\sin^2 4x = 1$

$\sin 4x = \pm 1$

$$4x = \frac{\pi}{2} + 2k\pi \qquad \text{or} \qquad 4x = \frac{3\pi}{2} + 2k\pi$$

$$x = \frac{\pi}{8} + \frac{k\pi}{2} \qquad\qquad x = \frac{3\pi}{8} + \frac{k\pi}{2}$$

29. $\qquad 2 \sin^2 3\theta + \sin 3\theta - 1 = 0$
$(2 \sin 3\theta - 1)(\sin 3\theta + 1) = 0$
$2 \sin 3\theta - 1 = 0 \qquad\qquad\qquad \text{or} \quad \sin 3\theta + 1 = 0$
$\qquad 2 \sin 3\theta = 1 \qquad\qquad\qquad\qquad \sin 3\theta = -1$

$\qquad\qquad \sin 3\theta = \dfrac{1}{2} \qquad\qquad\qquad\qquad\qquad 3\theta = 270° + 360°k$
$\qquad\qquad\qquad\qquad\qquad\qquad\qquad\qquad\qquad\quad \theta = 90° + 120°k$

$3\theta = 30° + 360°k \quad \text{or} \quad 3\theta = 150° + 360°k$
$\theta = 10° + 120°k \qquad\qquad \theta = 50° + 120°k$

33. $\tan^2 3\theta = 3$
$\tan 3\theta = \pm\sqrt{3}$
$3\theta = 60° + 180°k \quad \text{or} \quad 3\theta = 120° + 180°k$
$\theta = 20° + 60°k \qquad\qquad \theta = 40° + 60°k$

37. $\qquad\qquad\qquad\qquad \sin\theta + \cos\theta = -1$
$\sin^2\theta + 2\sin\theta\cos\theta + \cos^2\theta = 1 \qquad\qquad$ Square both sides
$\qquad\qquad\qquad 1 + 2\sin\theta\cos\theta = 1 \qquad\qquad \sin^2\theta + \cos^2\theta = 1$
$\qquad\qquad\qquad\qquad\qquad \sin 2\theta = 0 \qquad\qquad \sin 2\theta = 2\sin\theta\cos\theta$
$\qquad\quad 2\theta = 0° + 360°k \quad \text{or} \quad 2\theta = 180° + 360°k$
$\qquad\qquad \theta = 180°k \qquad\qquad\qquad \theta = 90° + 180°k$

If we let k = 0 and 1, we get

$\theta = 0° \qquad\qquad \theta = 90°$
$\theta = 180° \qquad\qquad \theta = 90° + 180° = 270°$

Since we squared both sides, we must check:

$$\sin 0° + \cos 0° \overset{?}{=} -1 \qquad\qquad \sin 90° + \cos 90° \overset{?}{=} -1$$
$$0 + 1 \overset{?}{=} -1 \qquad\qquad\qquad 1 + 0 \overset{?}{=} -1$$
$$1 \neq -1 \qquad\qquad\qquad\qquad 1 \neq -1$$

$$\sin 180° + \cos 180° \overset{?}{=} -1 \qquad\qquad \sin 270° + \cos 270° \overset{?}{=} -1$$
$$0 + (-1) \overset{?}{=} -1 \qquad\qquad\qquad -1 + 0 \overset{?}{=} -1$$
$$-1 \overset{\checkmark}{=} -1 \qquad\qquad\qquad\qquad -1 \overset{\checkmark}{=} -1$$

The solutions are 180° and 270°.

41. $d = 10 \tan \pi t$ where $d = 10$
$$10 = 10 \tan \pi t$$
$$\tan \pi t = 1$$

$$\pi t = \frac{\pi}{4}$$

$$t = \frac{1}{4} \text{ second}$$

45. $\dfrac{\sin x}{1 + \cos x} = \dfrac{\sin x}{1 + \cos x} \cdot \dfrac{1 - \cos x}{1 - \cos x}$ Multiply by a fraction equal to one

$$= \frac{\sin x(1 - \cos x)}{1 - \cos^2 x} \qquad\qquad \text{Multiply}$$

$$= \frac{\sin x(1 - \cos x)}{\sin^2 x} \qquad\qquad \text{Pythagorean identity}$$

$$= \frac{1 - \cos x}{\sin x} \qquad\qquad\qquad \text{Reduce}$$

49. $\tan \dfrac{A}{2} = \dfrac{\sin \dfrac{A}{2}}{\cos \dfrac{A}{2}}$ Ratio identity

$$= \frac{\sqrt{\dfrac{1 - \cos A}{2}}}{\sqrt{\dfrac{1 + \cos A}{2}}} \qquad\qquad \text{Half-angle formulas}$$

$$= \frac{\sqrt{1 - \cos A}}{\sqrt{1 + \cos A}} \qquad\qquad \text{Divide}$$

$$= \frac{\sqrt{1 - \cos A}}{\sqrt{1 + \cos A}} \cdot \frac{\sqrt{1 + \cos A}}{\sqrt{1 + \cos A}} \qquad \text{Multiply by a fraction equal to one}$$

$$= \frac{\sqrt{1 - \cos^2 A}}{1 + \cos A} \qquad\qquad \text{Multiply}$$

114

$$= \frac{\sqrt{\sin^2 A}}{1 + \cos A} \qquad \qquad \text{Pythagorean identity}$$

$$= \frac{\sin A}{1 + \cos A} \qquad \qquad \text{Simplify}$$

53. If $90° \leqslant A \leqslant 180°$, then $45° \leqslant \frac{A}{2} \leqslant 90°$.

Also, if $\sin A = \frac{1}{3}$ and A is in Q II, then

$$\cos A = -\sqrt{1 - \sin^2 A}$$

$$= -\sqrt{1 - \frac{1}{9}} = -\sqrt{\frac{8}{9}} = -\frac{2\sqrt{2}}{3}$$

Therefore, $\cos \frac{A}{2} = \sqrt{\frac{1 + \cos A}{2}}$

$$= \sqrt{\frac{1 + (-\frac{2\sqrt{2}}{3})}{2}}$$

$$= \sqrt{\frac{3 - 2\sqrt{2}}{6}}$$

57. $\csc(A + B) = \frac{1}{\sin(A + B)}$

$$= \frac{1}{\sin A \cos B + \cos A \sin B}$$

If $\sin B = \frac{3}{5}$ with B in Q I, then

$$\cos B = \sqrt{1 - \sin^2 B}$$

$$= \sqrt{1 - \frac{9}{25}} = \sqrt{\frac{16}{25}} = \frac{4}{5}$$

From Problem 53, we have $\cos A = -\frac{2\sqrt{2}}{3}$.

Substituting these above, we get:

$$\csc(A + B) = \frac{1}{\frac{1}{3}(\frac{4}{5}) + (-\frac{2\sqrt{2}}{3})(\frac{3}{5})}$$

$$= \frac{1}{\frac{4}{15} - \frac{6\sqrt{2}}{15}}$$

$$= \frac{1}{\dfrac{4 - 6\sqrt{2}}{15}}$$

$$= \frac{15}{4 - 6\sqrt{2}}$$

Problem Set 6.4

1. $\sin t = x$ and $\cos t = y$.
    $\sin^2 t + \cos^2 t = 1$ Pythagorean identity
         $x^2 + y^2 = 1$ Substitute given values

    The graph is a circle with its center at the origin and $r = 1$.

5. $2 \sin t = x$ and $4 \cos t = y$

    $\sin t = \dfrac{x}{2}$ $\cos t = \dfrac{y}{4}$

    $\sin^2 t + \cos^2 t = 1$

       $(\dfrac{x}{2})^2 + (\dfrac{y}{4})^2 = 1$

         $\dfrac{x^2}{4} + \dfrac{y^2}{16} - 1$

    The graph is an ellipse with center at the origin. The intercepts on the major axis are $(0, 4)$ and $(0, -4)$. The intercepts on the minor axis are $(2, 0)$ and $(-2, 0)$.

9. $\sin t - 2 = x$ and $\cos t - 3 = y$
         $\sin t = x + 2$ $\cos t = y + 3$

        $\sin^2 t + \cos^2 t = 1$
    $(x + 2)^2 + (y + 3)^2 = 1$

    The graph is a circle with center at $(-2, -3)$ and $r = 1$.

13. $3 \cos t - 3 = x$ and $3 \sin t + 1 = y$

        $3 \cos t = x + 3$ $3 \sin t = y - 1$

          $\cos t = \dfrac{x + 3}{3}$ $\sin t = \dfrac{y - 1}{3}$

         $\cos^2 t + \sin^2 t = 1$

    $(\dfrac{x + 3}{3})^2 + (\dfrac{y - 1}{3})^2 = 1$

    $\dfrac{(x + 3)^2}{9} + \dfrac{(y - 1)^2}{9} = 1$

    $(x + 3)^2 + (y - 1)^2 = 9$

    The graph is a circle with center at $(-3, 1)$ and $r = 3$.

17. $3 \sec t = x$ and $3 \tan t = y$

$\sec t = \dfrac{x}{3}$ $\tan t = \dfrac{y}{3}$

$\tan^2 t + 1 = \sec^2 t$

$\left(\dfrac{y}{3}\right)^2 + 1 = \left(\dfrac{x}{3}\right)^2$

$1 = \dfrac{x^2}{9} - \dfrac{y^2}{9}$ or $\dfrac{x^2}{9} - \dfrac{y^2}{9} = 1$

21. $\cos 2t = x$ $\sin t = y$
     $2 \cos^2 t - 1 = x$ $\sin^2 t = y^2$
     $2 \cos^2 t = x + 1$

$\cos^2 t = \dfrac{x + 1}{2}$

$\cos^2 t + \sin^2 t = 1$ Pythagorean identity

$\dfrac{x + 1}{2} + y^2 = 1$ Substitute values from above

$x + 1 + 2y^2 = 2$ Multiply both sides by 2

$x = -2y^2 + 1$ Subtract $(x + 1)$ from both sides

25. $3 \sin t = x$ $2 \sin t = y$

$\sin t = \dfrac{x}{3}$ $\sin t = \dfrac{y}{2}$

$\sin t = \sin t$ Reflexive property of equality

$\dfrac{x}{3} = \dfrac{y}{2}$ Substitute values from above

$2x = 3y$ Multiply both sides by 6

29. Let $\alpha = \text{Tan}^{-1} \dfrac{1}{3}$ and $\beta = \text{Sin}^{-1} \dfrac{1}{4}$.

Then $\tan \alpha = \dfrac{1}{3}$ and $\sin \beta = \dfrac{1}{4}$ and α and β are in Q I.

We can draw a triangle for α and one for β:

Then we find the missing sides using the Pythagorean theorem.

From the figure to the right, we see that

$\sin \alpha = \dfrac{1}{\sqrt{10}}$ $\sin \beta = \dfrac{1}{4}$

$\cos \alpha = \dfrac{3}{\sqrt{10}}$ $\cos \beta = \dfrac{\sqrt{15}}{4}$

Therefore, $\sin(\alpha + \beta) = \sin \alpha \cos \beta + \cos \alpha \sin \beta$

$$= \frac{1}{\sqrt{10}}\left(\frac{\sqrt{15}}{4}\right) + \frac{3}{\sqrt{10}}\left(\frac{1}{4}\right)$$

$$= \frac{\sqrt{15}}{4\sqrt{10}} + \frac{3}{4\sqrt{10}}$$

$$= \frac{\sqrt{15} + 3}{4\sqrt{10}}$$

33. Let $\alpha = \text{Tan}^{-1} x$. Then $\tan \alpha = \frac{x}{1}$ and $-90° \leqslant \alpha \leqslant 90°$.

We can draw a triangle for α and label the sides.
Using the Pythagorean theorem, we find the hypotenuse:

$$\text{hypotenuse} = \sqrt{x^2 + 1^2} = \sqrt{x^2 + 1}$$

From the figure to the right, we see that $\sin \alpha = \dfrac{x}{\sqrt{x^2 + 1}}$

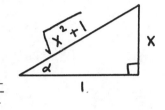

Therefore $\cos 2\alpha = 1 - 2 \sin^2 \alpha$

$$= 1 - 2\left(\frac{x}{\sqrt{x^2 + 1}}\right)^2$$

$$= 1 - 2\left(\frac{x^2}{x^2 + 1}\right)$$

$$= 1 - \frac{2x^2}{x^2 + 1}$$

$$= \frac{x^2 + 1 - 2x^2}{x^2 + 1}$$

$$= \frac{1 - x^2}{1 + x^2}$$

Chapter 6 Test

1. $2 \sin \theta - 1 = 0$

$\qquad 2 \sin \theta = 1$ — Add 1 to both sides

$\qquad \sin \theta = \dfrac{1}{2}$ — Divide both sides by 2

$\qquad \theta = 30° \text{ or } 150°$ — $\hat{\theta} = 30°$ and θ is in Q I or Q II

5. $4 \cos \theta - 2 \sec \theta = 0$

$\qquad 4 \cos \theta - \dfrac{2}{\cos \theta} = 0$ — $\sec \theta = \dfrac{1}{\cos \theta}$

$\qquad 4 \cos^2 \theta - 2 = 0$ — Multiply both sides by $\cos \theta$

$\qquad 4 \cos^2 \theta = 2$ — Add 2 to both sides

$\qquad \cos^2 \theta = \dfrac{1}{2}$ — Divide both sides by 4

$\qquad \cos \theta = \pm \dfrac{1}{\sqrt{2}}$ — Take square root of both sides

$\qquad \theta = 45°, 135°, 225°, 315°$ — $\hat{\theta} = 45°$

9. $\qquad 4 \cos 2\theta + 2 \sin \theta = 1$

$\qquad 4(1 - 2 \sin^2 \theta) + 2 \sin \theta = 1$ — $\cos 2\theta = 1 - 2 \sin^2 \theta$

$\qquad 4 - 8 \sin^2 \theta + 2 \sin \theta = 1$ — Simplify

$\qquad -8 \sin^2 \theta + 2 \sin \theta + 3 = 0$ — Subtract 1 from both sides

$\qquad 8 \sin^2 \theta - 2 \sin \theta - 3 = 0$ — Multiply both sides by -1

$\qquad (4 \sin \theta - 3)(2 \sin \theta + 1) = 0$ — Factor

$4 \sin \theta - 3 = 0 \qquad$ or $2 \sin \theta + 1 = 0$ — Set each factor = 0

$\qquad 4 \sin \theta = 3 \qquad\qquad 2 \sin \theta = -1$ — Solve each equation

$\qquad \sin \theta = \dfrac{3}{4} \qquad\qquad \sin \theta = -\dfrac{1}{2}$

$\qquad \theta = 48.6° \text{ or } 131.4° \qquad \theta = 210° \text{ or } 330°$

13. $\qquad \cos 2x - 3 \cos x = -2$

$\qquad 2 \cos^2 x - 1 - 3 \cos x = -2$ — $\cos 2x = 2 \cos^2 x - 1$

$\qquad 2 \cos^2 x - 3 \cos x + 1 = 0$ — Add 2 to both sides

$\qquad (2 \cos x - 1)(\cos x - 1) = 0$ — Factor

$2 \cos x - 1 = 0 \qquad$ or $\cos x - 1 = 0$ — Set each factor = 0

$\qquad 2 \cos x = 1 \qquad\qquad \cos x = 1$ — Solve each equation

$\qquad\qquad\qquad\qquad\qquad x = 0 + 2k\pi$

$\qquad \cos x = \dfrac{1}{2} \qquad\qquad x = 2k\pi$

$\qquad x = \dfrac{\pi}{3} + 2k\pi \text{ or}$

$\qquad\quad \dfrac{5\pi}{3} + 2k\pi$

17.
$$5 \sin^2 \theta - 3 \sin \theta = 2$$

$5 \sin^2 \theta - 3 \sin \theta - 2 = 0$	Subtract 2 from both sides
$(5 \sin \theta + 2)(\sin \theta - 1) = 0$	Factor

$5 \sin \theta + 2 = 0$ or $\sin \theta - 1 = 0$ Set each factor = 0

$\quad 5 \sin \theta = -2$ $\sin \theta = 1$ Solve each equation

$\qquad\qquad\qquad\qquad\qquad\qquad\qquad \theta = 90°$

$$\sin \theta = -\frac{2}{5}$$

$$\sin \theta = -0.4$$
$$\hat{\theta} = 23.6°$$
$$\theta = 203.6° \text{ or } 336.4°$$

21. $3 + 2 \sin t = x$ $1 + 2 \cos t = y$

$\qquad 2 \sin t = x - 3$ $2 \cos t = y - 1$

$$\sin t = \frac{x - 3}{2} \qquad\qquad \cos t = \frac{y - 1}{2}$$

$$\sin^2 t + \cos^2 t = 1$$

$$\left(\frac{x - 3}{2}\right)^2 + \left(\frac{y - 1}{2}\right)^2 = 1$$

$$\frac{(x - 3)^2}{4} + \frac{(y - 1)^2}{4} = 1$$

$$(x - 3)^2 + (y - 1)^2 = 4$$

The graph is a circle with center at (3, 1) and r = 2.

TRIANGLES

Problem Set 7.1

1. $\dfrac{b}{\sin B} = \dfrac{a}{\sin A}$ Law of Sines

$\dfrac{b}{\sin 60°} = \dfrac{12}{\sin 40°}$ Substitute given values

$b = \dfrac{12 \sin 60°}{\sin 40°}$ Multiply both sides by sin 60°

$= \dfrac{12(0.8660)}{0.6428}$ Calculator

$= 16$ cm   Round to 2 significant digits

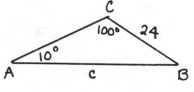

5. $\dfrac{c}{\sin C} = \dfrac{a}{\sin A}$ Law of Sines

$\dfrac{c}{\sin 100°} = \dfrac{24}{\sin 10°}$ Substitute given values

$c = \dfrac{24 \sin 100°}{\sin 10°}$ Multiply both sides by sin 100°

$= \dfrac{24(0.9848)}{0.1736}$ Calculator

$= 140$ yards   Round to 2 significant digits

9. $C = 180° - (A + B)$
    $= 180° - (52° + 48°)$
    $= 180° - 100°$
    $= 80°$

$\dfrac{a}{\sin A} = \dfrac{c}{\sin C}$ Law of Sines

$\dfrac{a}{\sin 52°} = \dfrac{14}{\sin 80°}$ Substitute given values

$a = \dfrac{14 \sin 52°}{\sin 80°}$ Multiply both sides by sin 52°

$= \dfrac{14(0.7880)}{0.9848}$ Calculator

$= 11$ cm.   Round to 2 significant digits

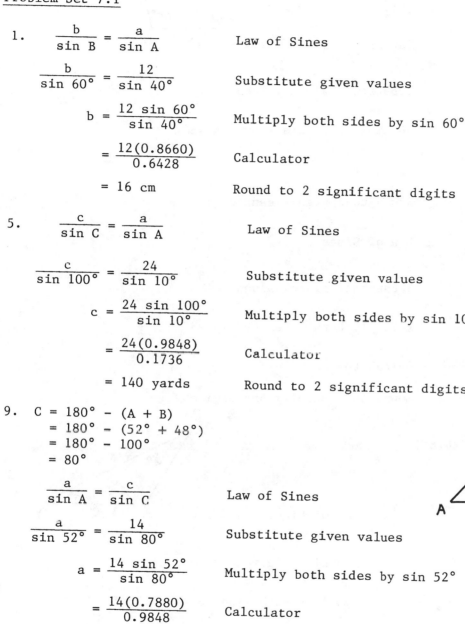

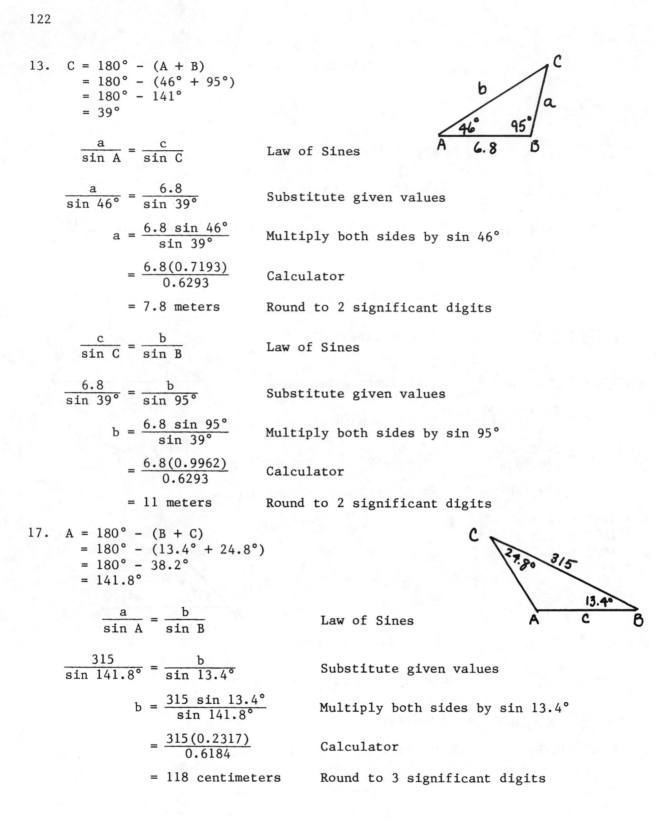

13. C = 180° − (A + B)
      = 180° − (46° + 95°)
      = 180° − 141°
      = 39°

$$\frac{a}{\sin A} = \frac{c}{\sin C}$$ Law of Sines

$$\frac{a}{\sin 46°} = \frac{6.8}{\sin 39°}$$ Substitute given values

$$a = \frac{6.8 \sin 46°}{\sin 39°}$$ Multiply both sides by sin 46°

$$= \frac{6.8(0.7193)}{0.6293}$$ Calculator

$$= 7.8 \text{ meters}$$ Round to 2 significant digits

$$\frac{c}{\sin C} = \frac{b}{\sin B}$$ Law of Sines

$$\frac{6.8}{\sin 39°} = \frac{b}{\sin 95°}$$ Substitute given values

$$b = \frac{6.8 \sin 95°}{\sin 39°}$$ Multiply both sides by sin 95°

$$= \frac{6.8(0.9962)}{0.6293}$$ Calculator

$$= 11 \text{ meters}$$ Round to 2 significant digits

17. A = 180° − (B + C)
      = 180° − (13.4° + 24.8°)
      = 180° − 38.2°
      = 141.8°

$$\frac{a}{\sin A} = \frac{b}{\sin B}$$ Law of Sines

$$\frac{315}{\sin 141.8°} = \frac{b}{\sin 13.4°}$$ Substitute given values

$$b = \frac{315 \sin 13.4°}{\sin 141.8°}$$ Multiply both sides by sin 13.4°

$$= \frac{315(0.2317)}{0.6184}$$ Calculator

$$= 118 \text{ centimeters}$$ Round to 3 significant digits

$$\frac{c}{\sin C} = \frac{a}{\sin A} \qquad \text{Law of Sines}$$

$$\frac{c}{\sin 24.8°} = \frac{315}{\sin 141.8°} \qquad \text{Substitute given value}$$

$$c = \frac{315(\sin 24.8°)}{\sin 141.8°} \qquad \text{Multiply both sides by sin 24.8°}$$

$$= \frac{315(0.4195)}{(0.6184)} \qquad \text{Calculator}$$

$$= 214 \text{ centimeters} \qquad \text{Round to 3 significant digits}$$

21. $\quad S = r\theta \;(\theta = \angle C) \qquad$ Arc length formula

$\qquad 11 = 12 \cdot \theta \qquad\qquad$ Substitute given values

$\qquad \theta = \dfrac{11}{12} \qquad\qquad$ Divide both sides by 12

We have that $\angle C = \dfrac{11}{12}$ radians. Converting this to degrees, we get:

$$\angle C = (\frac{11}{12} \cdot \frac{180}{\pi})° = 53°$$

$\angle D = 180° - (\angle C + \angle A)$
$\quad = 180° - (53° + 31°)$
$\quad = 180° - 84°$
$\quad = 96°$

Using the Law of Sines, we get

$$\frac{x + r}{\sin D} = \frac{r}{\sin A}$$

$$\frac{x + 12}{\sin 96°} = \frac{12}{\sin 31°}$$

$$x + 12 = \frac{12(\sin 96°)}{\sin 31°}$$

$$x + 12 = \frac{12(0.9945)}{0.5150}$$

$$x + 12 = 23$$

$$x = 11$$

25. We find the missing angles first.

$\qquad \angle ABD = 180° - 64° = 116°$

$\qquad \angle ADB = 180° - (46° + 116°) = 180° - 162° = 18°$

Now we find BD using the Law of Sines:

$$\frac{BD}{\sin A} = \frac{AB}{\sin ADB}$$

$$\frac{BD}{\sin 46°} = \frac{100}{\sin 18°}$$

$$BD = \frac{100 \sin 46°}{\sin 18°}$$

$$= \frac{100(0.7193)}{0.3090}$$

$$= 233$$

Then we find h, using the sine ratio:

$$\sin 64° = \frac{h}{233}$$

$$h = 233 \sin 64°$$
$$= 233(0.8988)$$
$$= 209 \text{ feet}$$

29.   First, we find the missing angles:

∠ADB = 90° − 35° = 55°

∠ADC = 180° − 55° = 125°

∠C   = 180° − (125° + 0.5°) = 180° − 125.5° = 54.5°

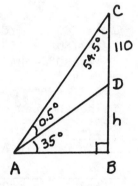

Next, we find AD using the Law of Sines:

$$\frac{AD}{\sin 54.5°} = \frac{110}{\sin 0.5°}$$

$$AD = \frac{110 \sin 54.5°}{\sin 0.5°}$$

$$= \frac{110(0.8141)}{0.0087}$$

$$= 10,262$$

Then, we find h using the sine ratio:

$$\sin 35° = \frac{h}{10,262}$$

$$h = 10,262 \sin 35°$$
$$= 10,262(0.5736)$$
$$= 5900 \text{ feet}$$

33. First, we find the missing angle:

$$\angle C = 180° - (53° + 31°)$$
$$= 180° - 84°$$
$$= 96°$$

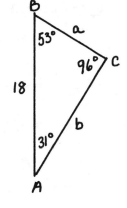

Then we find the missing sides using the Law of Sines:

$$\frac{18}{\sin 96°} = \frac{a}{\sin 31°}$$

$$a = \frac{18 \sin 31°}{\sin 96°}$$

$$= \frac{18(0.5150)}{0.9945}$$

$$= 9.3 \text{ miles}$$

$$\frac{18}{\sin 96°} = \frac{b}{\sin 53°}$$

$$b = \frac{18 \sin 53°}{\sin 96°}$$

$$= \frac{18(0.7986)}{0.9945}$$

$$= 14 \text{ miles}$$

37. $\sin \theta \cos \theta - 2 \cos \theta = 0$

$\cos \theta(\sin \theta - 2) = 0$ Factor

$\cos \theta = 0$ or $\sin \theta - 2 = 0$ Set each factor = 0

$\theta = 90°$ or $270°$ $\sin \theta = 2$ Solve each equation

              No solution

41. $\cos^2 \theta - 4 \cos \theta + 2 = 0$ Standard form

$a = 1$, $b = -4$, and $c = 2$

$$\cos \theta = \frac{-(-4) \pm \sqrt{(-4)^2 - 4(1)(2)}}{2(1)} \quad \text{Quadratic formula}$$

$$= \frac{4 \pm \sqrt{8}}{2} \quad \text{Simplify}$$

$$= \frac{4 \pm 2\sqrt{2}}{2}$$

$$= 2 \pm \sqrt{2}$$

Then $\cos \theta = 2 + 1.4142$ or $\cos \theta = 2 - 1.4142$

       $\cos \theta = 3.4142$ $\cos \theta = 0.5858$

        No solution $\theta = 54.1°$ or $305.9°$

45. $\sin \theta = 0.7380$

    $\theta = 47.6°$ or $132.4°$ $\hat{\theta} = 47.6°$ and θ is in Q I or Q II

Problem Set 7.2

1. $\sin B = \dfrac{b \sin A}{a}$

    $= \dfrac{40 \sin 30°}{10}$

    $= \dfrac{40(0.5)}{10}$

    $= 2$

    Since sin B can never be greater than 1, no triangle exists.

5. $\sin B = \dfrac{b \sin A}{a}$

    $= \dfrac{18 \sin 60°}{16}$

    $= \dfrac{18(0.8660)}{16}$

    $= 0.9743$

    $B = 77°$ or $B' = 180° - 77°$
    $B' = 103°$

    Therefore, there are 2 possible triangles.

9. $\sin B = \dfrac{b \sin A}{a}$

    $= \dfrac{22.3 \sin 112.2°}{43.8}$

    $= \dfrac{22.3(0.9259)}{43.8}$

    $= 0.4714$

    $B = 28.1°$ or $B' = 180° - 28.1°$
    $= \cancel{151.9°}$

    B' cannot equal 151.9° because A = 112.2°.
    There can only be one obtuse angle in a triangle.

    $C = 180° - (A + B)$
    $= 180° - (112.2° + 28.1°)$
    $= 180° - 140.3°$
    $= 39.7°$

$$\frac{c}{\sin C} = \frac{a}{\sin A}$$

$$c = \frac{a \sin C}{\sin A}$$

$$= \frac{43.8 \sin 39.7°}{\sin 112.2°}$$

$$= \frac{43.8(0.6388)}{0.9259}$$

$$= 30.2 \text{ cm}$$

13. $\sin C = \dfrac{c \sin B}{b}$

$$= \frac{1.12 \sin 45°10'}{1.79}$$

$$= \frac{1.12(0.7092)}{1.79}$$

$$= 0.4437$$

$C = 26.3°$ or $C' = 180° - 26.3°$
$C = 26°20'$ $C' = \cancel{153.7°}$

$C' \neq 153.7°$ because $B + C' = 45.2° + 153.7°$
$\qquad\qquad\qquad\qquad\qquad\quad = 198.9°$ This is impossible.

$A = 180° - (B + C)$
$\;\; = 180° - (45°10' + 26°20')$
$\;\; = 180° - 71°30'$
$\;\; = 108°30'$

$$\frac{a}{\sin A} = \frac{b}{\sin B}$$

$$a = \frac{b \sin A}{\sin B}$$

$$= \frac{1.79 \sin 108°30'}{\sin 45°10'}$$

$$= \frac{1.79(0.9483)}{0.7092}$$

$$= 2.39 \text{ inches}$$

17. $\sin B = \dfrac{b \sin A}{a}$

$$= \frac{2.9 \sin 142°}{1.4}$$

$$= \frac{2.9(0.6157)}{1.4}$$

$$= 1.2753$$

Since sin B can never be greater than 1, no triangle exists.

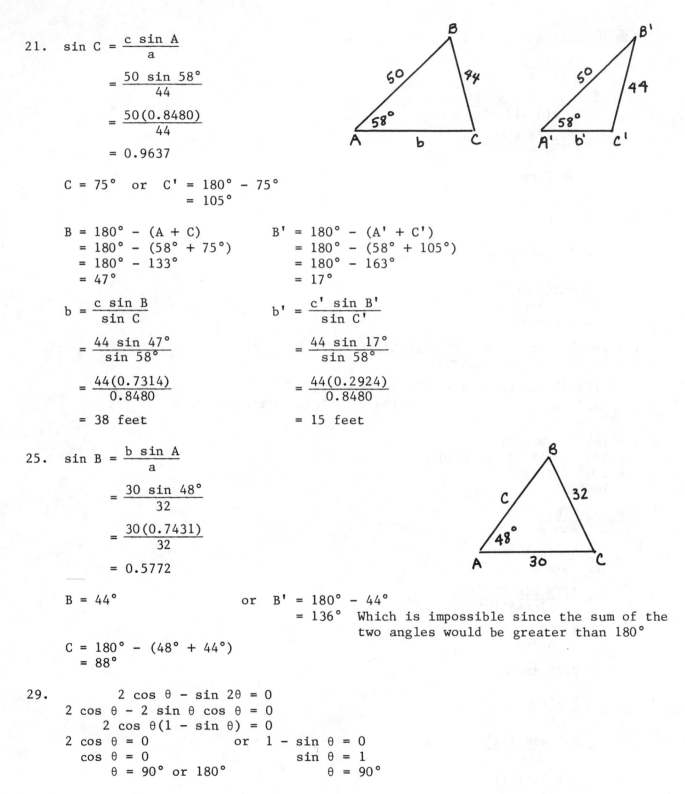

21. $\sin C = \dfrac{c \sin A}{a}$

$= \dfrac{50 \sin 58°}{44}$

$= \dfrac{50(0.8480)}{44}$

$= 0.9637$

$C = 75°$ or $C' = 180° - 75°$
$\qquad\qquad\qquad = 105°$

$B = 180° - (A + C)$ $B' = 180° - (A' + C')$
$ = 180° - (58° + 75°)$ $= 180° - (58° + 105°)$
$ = 180° - 133°$ $= 180° - 163°$
$ = 47°$ $= 17°$

$b = \dfrac{c \sin B}{\sin C}$ $b' = \dfrac{c' \sin B'}{\sin C'}$

$= \dfrac{44 \sin 47°}{\sin 58°}$ $= \dfrac{44 \sin 17°}{\sin 58°}$

$= \dfrac{44(0.7314)}{0.8480}$ $= \dfrac{44(0.2924)}{0.8480}$

$= 38$ feet $= 15$ feet

25. $\sin B = \dfrac{b \sin A}{a}$

$= \dfrac{30 \sin 48°}{32}$

$= \dfrac{30(0.7431)}{32}$

$= 0.5772$

$B = 44°$ or $B' = 180° - 44°$
$\qquad\qquad\qquad\qquad\qquad = 136°$ Which is impossible since the sum of the
$\qquad\qquad\qquad\qquad\qquad\qquad$ two angles would be greater than 180°

$C = 180° - (48° + 44°)$
$ = 88°$

29. $\qquad 2 \cos \theta - \sin 2\theta = 0$
$2 \cos \theta - 2 \sin \theta \cos \theta = 0$
$\qquad 2 \cos \theta(1 - \sin \theta) = 0$
$2 \cos \theta = 0 \qquad\qquad$ or $\quad 1 - \sin \theta = 0$
$\quad \cos \theta = 0 \qquad\qquad\qquad\qquad \sin \theta = 1$
$\qquad\quad \theta = 90°$ or $180° \qquad\qquad\quad \theta = 90°$

33. $2 \cos x - \sec x + \tan x = 0$

$$2 \cos x - \frac{1}{\cos x} + \frac{\sin x}{\cos x} = 0$$

$$2 \cos^2 x - 1 + \sin x = 0$$
$$2(1 - \sin^2 x) - 1 + \sin x = 0$$
$$2 - 2 \sin^2 x - 1 + \sin x = 0$$
$$2 \sin^2 x - \sin x - 1 = 0$$
$$(2 \sin x + 1)(\sin x - 1) = 0$$

$2 \sin x + 1 = 0$ or $\sin x - 1 = 0$
   $2 \sin x = -1$ $\sin x = 1$

$$\sin x = -\frac{1}{2} \qquad\qquad\qquad x = \frac{\pi}{2} + 2k\pi$$

$$x = \frac{7\pi}{6} + 2k\pi \text{ or } \frac{11\pi}{6} + 2k\pi$$

Problem Set 7.3

1. $c^2 = a^2 + b^2 - 2ab \cos C$
   $= (120)^2 + (66)^2 - 2(120)(66) \cos 60°$
   $= 14{,}400 + 4356 - 15{,}840(0.5)$
   $= 14{,}400 + 4356 - 7{,}920$
   $= 10{,}836$

   $c = 104$
   $= 100$ inches (rounded to 2 significant digits)

5. $a^2 = b^2 + c^2 - 2bc \cos A$
   $= (4.2)^2 + (6.8)^2 - 2(4.2)(6.8) \cos 116°$
   $= 17.64 + 46.24 - 57.12(-0.4384)$
   $= 17.64 + 46.24 + 25.04$
   $= 88.92$

   $a = 9.4$ meters

9. $b^2 = a^2 + c^2 - 2ac \cos B$
   $= (410)^2 + (340)^2 - 2(410)(340) \cos 151.5°$
   $= 168{,}100 + 115{,}600 - 278{,}800(-0.8788)$
   $= 528{,}714.211$

   $b = 727$ meters

$$\sin A = \frac{a \sin B}{b}$$

$$= \frac{410 \sin 151.5°}{727}$$

$$= \frac{410(0.4772)}{727}$$

$$= 0.2691$$

$$A = 15.6°$$

$$C = 180° - (A + B)$$
$$= 180° - (15.6° + 151.5°)$$
$$= 180° - (167.1°)$$
$$= 12.9°$$

13. $a^2 = b^2 + c^2 - 2bc \cos A$
$$= (0.923)^2 + (0.387)^2 - 2(0.923)(0.387) \cos 43°20'$$
$$= 0.851929 + 0.149769 - 0.714402(0.7274)$$
$$= 0.4821$$

$a = 0.694$ kilometers

$$\sin C = \frac{c \sin A}{a}$$

$$= \frac{0.387 \sin 43°20'}{0.694}$$

$$= \frac{0.387(0.6862)}{0.694}$$

$$= 0.3827$$

$C = 22°30'$

$$B = 180° - (A + C)$$
$$= 180° - (22°30' + 43°20')$$
$$= 180° - (65°50')$$
$$= 114°10'$$

17. $a^2 = b^2 + c^2 - 2bc \cos A$
$a^2 = b^2 + c^2 - 2bc(\cos 90°)$
$a^2 = b^2 + c^2 - 2bc(0)$
$a^2 = b^2 + c^2$

21. $d_1 = r_1 t_1$ $d_2 = r_2 t_2$
$\quad = 130(1.5)$ $\quad = 150(1.5)$
$\quad = 195$ miles $\quad = 225$ miles

$a^2 = b^2 + c^2 - 2bc \cos A$
$$= (225)^2 + (195)^2 - 2(225)(195) \cos 36°$$
$$= 17658.76$$

$a = 133$
$a = 130$ miles (rounded to 2 significant digits)

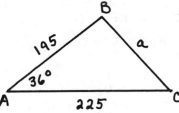

25. $|\vec{v} + \vec{w}|^2 = |\vec{v}|^2 + |\vec{w}|^2 - 2|\vec{v}||\vec{w}| \cos \theta$
$$= (35)^2 + (160)^2 - 2(35)(160) \cos 165°$$
$$= 1225 + 25{,}600 - 11{,}200(-0.9659)$$
$$= 1225 + 25{,}600 + 10818$$
$$= 37{,}643$$
$$= 190 \text{ mph (to two significant digits)}$$

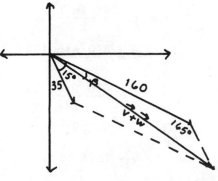

$$\frac{\sin \beta}{35} = \frac{\sin 165°}{190}$$

$$\sin \beta = \frac{35 \sin 165°}{190}$$

$$= \frac{35(0.2588)}{190}$$

$$\sin \beta = 0.0477$$

$$\beta = 3° \text{ (to the nearest degree)}$$

The true course is $150° + \beta = 150° + 3° = 153°$. The speed of the plane with respect to the ground is 190 miles per hour.

29. $\sin 3x = \frac{1}{2}$

$$3x = \frac{\pi}{6} + 2k\pi \quad \text{or} \quad 3x = \frac{5\pi}{6} + 2k\pi$$

$$x = \frac{\pi}{18} + \frac{2k\pi}{3} \qquad x = \frac{5\pi}{18} + \frac{2k\pi}{3}$$

33. $2 \cos^2 3\theta - 9 \cos 3\theta + 4 = 0$
$(2 \cos 3\theta - 1)(\cos 3\theta - 4) = 0$
$2 \cos 3\theta - 1 = 0 \quad \text{or} \quad \cos 3\theta - 4 = 0$
$\quad 2 \cos 3\theta = 1 \qquad\qquad \cos 3\theta = 4$
$$\qquad\qquad\qquad\qquad\qquad \text{No solution}$$
$$\cos 3\theta = \frac{1}{2}$$

$3\theta = 60° + 360°k \quad \text{or} \quad 3\theta = 300° + 360°k$
$\theta = 20° + 120°k \quad \text{or} \quad \theta = 100° + 120°k$

37.
$$\sin \theta + \cos \theta = 1$$
$$(\sin \theta + \cos \theta)^2 = 1^2$$
$$\sin^2 \theta + 2 \sin \theta \cos \theta + \cos^2 \theta = 1$$
$$(\sin^2 \theta + \cos^2 \theta) + 2 \sin \theta \cos \theta = 1$$
$$1 + \sin 2\theta = 1$$
$$\sin 2\theta = 0$$
$$2\theta = 0° \text{ or } 180°$$
$$\theta = 0° \text{ or } 90°$$

Check: $\sin 0° + \cos 0° = 1 \qquad\qquad \sin 90° + \cos 90° = 1$
$\qquad\qquad\qquad 0 + 1 \overset{\checkmark}{=} 1 \qquad\qquad\qquad\qquad 1 + 0 \overset{\checkmark}{=} 1$

Problem Set 7.4

1. $S = \frac{1}{2} ab \sin C$ Formula for area of a triangle

$\quad = \frac{1}{2} (50)(70) \sin 60°$ Substitute known values

$\quad = 1750(0.8660)$ Simplify

$\quad = 1520 \text{ cm}^2$ Rounded to 2 significant digits

5. $S = \frac{1}{2} bc \sin A$ Formula for area of a triangle

      $= \frac{1}{2} (0.923)(0.387) \sin 43°20'$ Substitute known values

      $= (0.1786)(0.6862)$ Simplify

      $= 0.123 \text{ km}^2$ Round to 3 significant digits

9. $C = 180° - (A + B)$
      $= 180° - (42.5° + 71.4°)$
      $= 180° - 113.9°$
      $= 66.1°$

      $S = \dfrac{a^2 \sin B \sin C}{2 \sin A}$ Formula for area of a triangle

      $= \dfrac{(210)^2 (\sin 71.4°)(\sin 66.1°)}{2 \sin 42.5°}$ Substitute known values

      $= \dfrac{(44,100)(0.9478)(0.9143)}{2(0.6756)}$ Simplify

      $= 28,300 \text{ in}^2$ Round to 3 significant digits

13. $s = \frac{1}{2} (a + b + c)$ Formula for half perimeter

      $= \frac{1}{2} (44 + 66 + 88)$ Substitute known values

      $= 99$ Simplify

      $S = \sqrt{s(s - a)(s - b)(s - c)}$ Formula for area of a triangle
      $= \sqrt{99(99 - 44)(99 - 66)(99 - 88)}$ Substitute known values
      $= \sqrt{99(55)(33)(11)}$ Simplify
      $= \sqrt{1,976,535}$
      $= 1410 \text{ in}^2$ Round to 3 significant digits

17. $s = \frac{1}{2} (a + b + c)$ (Explanation same as #13)

      $= \frac{1}{2} (4.38 + 3.79 + 5.22)$

      $= 6.695$

      $S = \sqrt{s(s - a)(s - b)(s - c)}$
      $= \sqrt{6.695(6.695 - 4.38)(6.695 - 3.79)(6.695 - 5.22)}$
      $= \sqrt{6.695(2.315)(2.905)(1.475)}$
      $= \sqrt{66.41}$
      $= 8.15 \text{ ft}^2$

21. $C = 180° - (A + B)$
$= 180° - (30° + 50°)$
$= 180° - 80°$
$= 100°$

$S = \dfrac{c^2 \sin A \sin B}{2 \sin C}$ Formula for area of a triangle

$c^2 = \dfrac{2S \sin C}{\sin A \sin B}$ Multiply both sides by $\dfrac{2 \sin C}{\sin A \sin B}$

$= \dfrac{2(40) \sin 100°}{\sin 30° \sin 50°}$ Substitute known values

$= \dfrac{80(0.9848)}{0.5(0.7660)}$ Simplify

$c^2 = 205.69$

$c = 14.3$ cm   Round to 3 significant digits

25. $3 + 2 \sin t = x$ $1 + 2 \cos t = y$

     $2 \sin t = x - 3$ $2 \cos t = y - 1$

       $\sin t = \dfrac{x - 3}{2}$ $\cos t = \dfrac{y - 1}{2}$

$\sin^2 t + \cos^2 t = 1$

$\left(\dfrac{x - 3}{2}\right)^2 + \left(\dfrac{y - 1}{2}\right)^2 = 1$

$\dfrac{(x - 3)^2}{4} + \dfrac{(y - 1)^2}{4} = 1$

$(x - 3)^2 + (y - 1)^2 = 4$

The graph is a circle with center at (3, 1) and radius of 2.

29. $\cos 2t = y$ and $\sin t = x$
$\cos 2t = 1 - \sin^2 t$
$y = 1 - 2x^2$

Chapter 7 Test

1. $b = \dfrac{a \sin B}{\sin A}$

    $= \dfrac{3.8 \sin 70°}{\sin 32°}$

    $= \dfrac{3.8(0.9397)}{0.5299}$

    $= 6.7$ in.

5. $\sin B = \dfrac{b \sin A}{a}$

    $= \dfrac{42 \sin 60°}{12}$

    $= \dfrac{42(0.8660)}{12}$

    $= 3.0311$

Since sin B is never greater than 1, no triangle exists.

9. $c^2 = a^2 + b^2 - 2ab \cos C$
    $= (10)^2 + (12)^2 - 2(10)(12) \cos 60°$
    $= 100 + 144 - 240(0.5)$
    $= 124$

    $c = 11$ cm

13. $c^2 = a^2 + b^2 - 2ab \cos C$
    $= (6.4)^2 + (2.8)^2 - 2(6.4)(2.8) \cos 119°$
    $= 40.96 + 7.84 - 35.84(-0.4848)$
    $= 66.18$

    $c = 8.1$ cm

    $\sin B = \dfrac{b \sin C}{c}$

    $= \dfrac{2.8 \sin 119°}{8.1}$

    $= \dfrac{2.8(0.8746)}{8.1}$

    $= 0.3023$

    $B = 18°$

    $A = 180° - (B + C)$
    $= 180° - (18° + 119°)$
    $= 180° - 137°$
    $= 43°$

17. First, we find the missing angles of △ABD:

$$\angle ABD = 180° - 64° = 116°$$

$$\angle ADB = 180° - (43° + 116°) = 21°$$

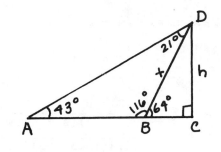

Next, we find x using the Law of Sines:

$$\frac{x}{\sin 43°} = \frac{240}{\sin 21°}$$

$$x = \frac{240 \sin 43°}{\sin 21°}$$

$$= \frac{240(0.6820)}{0.3584}$$

$$= 457$$

Then, we find h using the sine relationship:

$$\sin 64° = \frac{h}{x}$$

$$h = x \sin 64°$$
$$= 457(0.8988)$$
$$= 410 \text{ ft} \quad \text{(rounded to 2 significant digits)}$$

21. C = 180° - (A + B)
   = 180° - (47° + 37°)
   = 180° - 84°
   = 96°

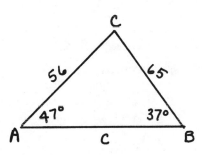

We find c using the Law of Cosines:

$$c^2 = a^2 + b^2 - 2ab \cos C$$
$$= (56)^2 + (65)^2 - 2(56)(65) \cos 96°$$
$$= 3136 + 4225 - 7280(-0.1045)$$
$$= 8122$$

$$c = 90 \text{ ft}$$

25. C = 180° - (A + B)
   = 180° - (38.2° + 63.4°)
   = 180° - 101.6°
   = 78.4°

$$S = \frac{c^2 \sin A \sin B}{2 \sin C}$$

$$= \frac{(42.0)^2 \sin 38.2° \sin 63.4°}{2 \sin 78.4°}$$

$$= \frac{1764(0.6184)(0.8942)}{2(0.9796)}$$

$$= 498 \text{ cm}^2$$

29. $s = \frac{1}{2}(a + b + c)$

$= \frac{1}{2}(5 + 7 + 9)$

$= 10.5$

$S = \sqrt{s(s - a)(s - b)(s - c)}$

$= \sqrt{10.5(10.5 - 5)(10.5 - 7)(10.5 - 9)}$

$= \sqrt{10.5(5.5)(3.5)(1.5)}$

$= \sqrt{303.1875}$

$= 17 \text{ km}^2$

COMPLEX NUMBERS AND POLAR COORDINATES

<u>Problem Set 8.1</u>

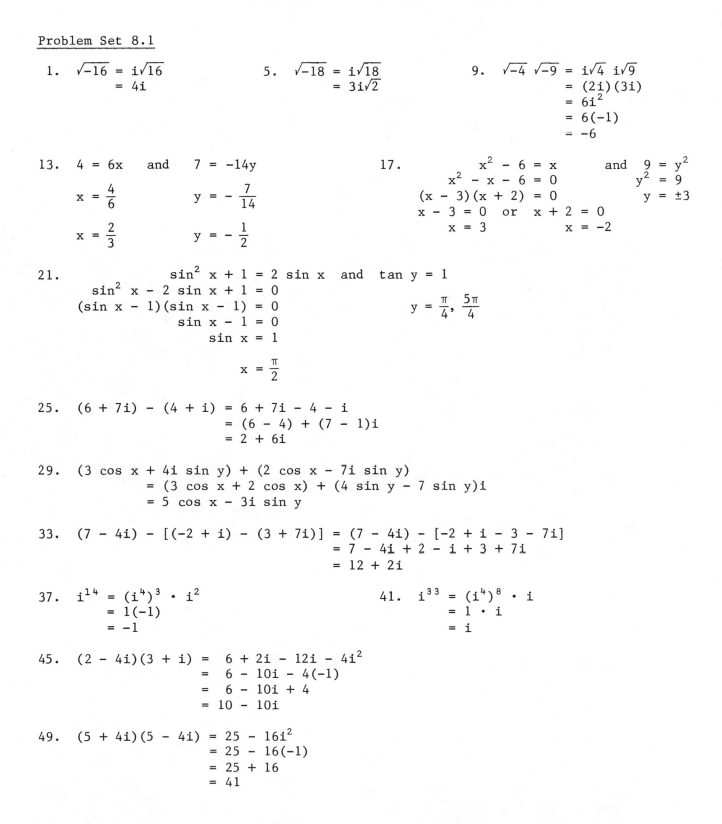

1. $\sqrt{-16} = i\sqrt{16}$
    $= 4i$

5. $\sqrt{-18} = i\sqrt{18}$
    $= 3i\sqrt{2}$

9. $\sqrt{-4}\ \sqrt{-9} = i\sqrt{4}\ i\sqrt{9}$
    $= (2i)(3i)$
    $= 6i^2$
    $= 6(-1)$
    $= -6$

13. $4 = 6x$ and $7 = -14y$

    $x = \dfrac{4}{6}$ $y = -\dfrac{7}{14}$

    $x = \dfrac{2}{3}$ $y = -\dfrac{1}{2}$

17. $x^2 - 6 = x$ and $9 = y^2$
         $x^2 - x - 6 = 0$ $y^2 = 9$
         $(x - 3)(x + 2) = 0$ $y = \pm 3$
         $x - 3 = 0$ or $x + 2 = 0$
             $x = 3$ $x = -2$

21. $\sin^2 x + 1 = 2 \sin x$ and $\tan y = 1$
         $\sin^2 x - 2 \sin x + 1 = 0$
       $(\sin x - 1)(\sin x - 1) = 0$ $y = \dfrac{\pi}{4}, \dfrac{5\pi}{4}$
                 $\sin x - 1 = 0$
                     $\sin x = 1$

                         $x = \dfrac{\pi}{2}$

25. $(6 + 7i) - (4 + i) = 6 + 7i - 4 - i$
    $= (6 - 4) + (7 - 1)i$
    $= 2 + 6i$

29. $(3 \cos x + 4i \sin y) + (2 \cos x - 7i \sin y)$
    $= (3 \cos x + 2 \cos x) + (4 \sin y - 7 \sin y)i$
    $= 5 \cos x - 3i \sin y$

33. $(7 - 4i) - [(-2 + i) - (3 + 7i)] = (7 - 4i) - [-2 + i - 3 - 7i]$
    $= 7 - 4i + 2 - i + 3 + 7i$
    $= 12 + 2i$

37. $i^{14} = (i^4)^3 \cdot i^2$
    $= 1(-1)$
    $= -1$

41. $i^{33} = (i^4)^8 \cdot i$
    $= 1 \cdot i$
    $= i$

45. $(2 - 4i)(3 + i) = 6 + 2i - 12i - 4i^2$
    $= 6 - 10i - 4(-1)$
    $= 6 - 10i + 4$
    $= 10 - 10i$

49. $(5 + 4i)(5 - 4i) = 25 - 16i^2$
    $= 25 - 16(-1)$
    $= 25 + 16$
    $= 41$

53. $2i(3 + i)(2 + 4i) = 2i(6 + 12i + 2i + 4i^2)$
$\qquad\qquad\qquad\quad = 2i(6 + 14i - 4)$
$\qquad\qquad\qquad\quad = 2i(2 + 14i)$
$\qquad\qquad\qquad\quad = 4i + 28i^2$
$\qquad\qquad\qquad\quad = -28 + 4i$

57. $\dfrac{2i}{3 + i} \cdot \dfrac{3 - i}{3 - i} = \dfrac{6i - 2i^2}{9 - i^2}$

$\qquad\qquad\qquad = \dfrac{6i - 2(-1)}{9 - (-1)}$

$\qquad\qquad\qquad = \dfrac{2 + 6i}{10}$

$\qquad\qquad\qquad = \dfrac{1 + 3i}{5}$

$\qquad\qquad\qquad = \dfrac{1}{5} + \dfrac{3}{5}i$

61. $\dfrac{5 - 2i}{i} \cdot \dfrac{i}{i} = \dfrac{5i - 2i^2}{i^2}$

$\qquad\qquad\qquad = \dfrac{5i - 2(-1)}{-1}$

$\qquad\qquad\qquad = \dfrac{2 + 5i}{-1}$

$\qquad\qquad\qquad = -2 - 5i$

65. $z_1 z_2 = (2 + 3i)(2 - 3i)$
$\qquad\quad = 4 - 9i^2$
$\qquad\quad = 4 - 9(-1)$
$\qquad\quad = 4 + 9$
$\qquad\quad = 13$

69. $2z_1 + 3z_2 = 2(2 + 3i) + 3(2 - 3i)$
$\qquad\qquad\quad = 4 + 6i + 6 - 9i$
$\qquad\qquad\quad = 10 - 3i$

73. $(x + 3i)(x - 3i) = x^2 - 9i^2$
$\qquad\qquad\qquad\quad = x^2 - 9(-1)$
$\qquad\qquad\qquad\quad = x^2 + 9$

81. $\qquad r = \sqrt{a^2 + b^2}$

$\sin \theta = \dfrac{b}{\sqrt{a^2 + b^2}} \quad \text{and} \quad \cos \theta = \dfrac{a}{\sqrt{a^2 + b^2}}$

85. $a^2 = b^2 + c^2 - 2bc \cos A$
$\quad = (243)^2 + (157)^2 - 2(243)(157) \cos 73.1°$
$\quad = 59{,}049 + 24{,}649 - 76{,}302(0.2907)$
$\quad = 61{,}517$

$a = 248$ cm

$\sin C = \dfrac{c \sin A}{a}$

$\qquad = \dfrac{157 \sin 73.1°}{248}$

$\qquad = \dfrac{157(0.9568)}{248}$

$\qquad = 0.6057$

$\quad C = 37.2°$

$B = 180° - (A + C)$
$\quad = 180° - (73.1° + 37.2°)$
$\quad = 180° - 110.3°$
$\quad = 69.7°$

Problem Set 8.2

1. $|3 + 4i| = \sqrt{3^2 + 4^2}$
   $= \sqrt{9 + 16}$
   $= \sqrt{25}$
   $= 5$

5. $|0 - 5i| = \sqrt{0^2 + (-5)^2}$
   $= \sqrt{25}$
   $= 5$

9. $|-4 - 3i| = \sqrt{(-4)^2 + (-3)^2}$
   $= \sqrt{16 + 9}$
   $= \sqrt{25}$
   $= 5$

13. Opposite of $4i = -4i$
    Conjugate of $4i = -4i$

17. Opposite of $-5 - 2i = -(-5 - 2i)$
    $= 5 + 2i$

    Conjugate of $-5 - 2i = -5 + 2i$

21. $4(\cos 120° + i \sin 120°) = 4[-\frac{1}{2} + i(\frac{\sqrt{3}}{2})]$
    $= -2 + 2i\sqrt{3}$

25. $\cos 315° + i \sin 315° = \frac{\sqrt{2}}{2} + i(-\frac{\sqrt{2}}{2})$
    $= \frac{\sqrt{2}}{2} - \frac{\sqrt{2}}{2} i$

29. $100(\cos 143° + i \sin 143°) = 100[-0.7986 + i(0.6018)]$
    $= -79.86 + 60.18i$

33. $10(\cos 342° + i \sin 342°) = 10[0.951 + i(-0.309)]$
    $= 9.51 - 3.09i$

37. We have $x = 1$ and $y = -1$, therefore
    $r = \sqrt{1^2 + (-1)^2} = \sqrt{2}$

    We also know that $\tan \theta = \frac{-1}{1} = -1$ and θ is in Q IV.

    Therefore, $\theta = 315°$.

    In trigonometric form, $z = r(\cos \theta + i \sin \theta)$
    $= \sqrt{2}(\cos 315° + i \sin 315°)$

140

41. We have x = 0 and y = 8, therefore

$$r = \sqrt{0^2 + 8^2} = 8$$

We also know that θ = 90°.

In trigonometric form, z = r(cos θ + i sin θ)
$$= 8(\cos 90° + i \sin 90°)$$

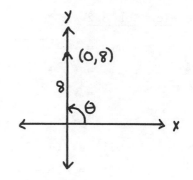

45. We have x = -2 and y = 2√3, therefore

$$r = \sqrt{(-2)^2 + (2\sqrt{3})^2}$$
$$= \sqrt{4 + 12}$$
$$= \sqrt{16}$$
$$= 4$$

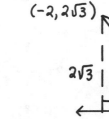

We also know that $\tan θ = \dfrac{2\sqrt{3}}{-2} = -\sqrt{3}$ and θ is in Q II. Therefore, θ = 120°.

In trigonometric form, z = r(cos θ + i sin θ)
$$= 4(\cos 120° + i \sin 120°)$$

49. $\quad 2(\cos 30° + i \sin 30°) = 2(\dfrac{\sqrt{3}}{2} + i \cdot \dfrac{1}{2}) = \sqrt{3} + i$

$\quad 2[\cos(-30°) + i \sin(-30°)] = 2[\dfrac{\sqrt{3}}{2} + i \cdot (-\dfrac{1}{2})] = \sqrt{3} - i$

53. cos 75° = cos (30° + 45°)

$$= \cos 30° \cos 45° - \sin 30° \sin 45°$$
$$= \frac{\sqrt{3}}{2} \cdot \frac{\sqrt{2}}{2} - \frac{1}{2} \cdot \frac{\sqrt{2}}{2}$$
$$= \frac{\sqrt{6}}{4} - \frac{\sqrt{2}}{4}$$
$$= \frac{\sqrt{6} - \sqrt{2}}{4}$$

57. sin 30° cos 90° + cos 30° sin 90° = sin(30° + 90°)

$$= \sin 120°$$
$$= \frac{\sqrt{3}}{2}$$

61. $\sin B = \dfrac{b \sin A}{a}$

   $\qquad = \dfrac{567 \sin 45.6°}{234}$

   $\qquad = \dfrac{567(0.7145)}{234}$

   $\qquad = 1.7312$

Since sin B is never greater than 1, no triangle exists.

Problem Set 8.3

1. $3(\cos 20° + i \sin 20°) \cdot 4(\cos 30° + i \sin 30°)$
   $\qquad = 3 \cdot 4[\cos(20° + 30°) + i \sin(20° + 30°)]$
   $\qquad = 12(\cos 50° + i \sin 50°)$

5. $2(\cos 135° + i \sin 135°) \cdot 2(\cos 45° + i \sin 45°)$
   $\qquad = 2 \cdot 2[\cos(135° + 45°) + i \sin(135° + 45°)]$
   $\qquad = 4(\cos 180° + i \sin 180°)$

9. $z_1 z_2 = (1 + i\sqrt{3})(-\sqrt{3} + i)$
   $\qquad = -\sqrt{3} + i - 3i + i^2\sqrt{3}$
   $\qquad = -\sqrt{3} - 2i - \sqrt{3}$
   $\qquad = -2\sqrt{3} - 2i$

   $z_1 = 1 + i\sqrt{3}$ where $x = 1$, $y = \sqrt{3}$, and $r = \sqrt{1^2 + (\sqrt{3})^2} = 2$.

   Also, $\tan \theta = \sqrt{3}$ and θ is in Q I.
   Therefore, $\theta = 60°$.

   $z_1 = 2(\cos 60° + i \sin 60°)$ in trigonometric form

   $z_2 = -\sqrt{3} + i$ where $x = -\sqrt{3}$, $y = 1$, and $r = \sqrt{(-\sqrt{3})^2 + 1^2} = 2$.

   Also, $\tan \theta = -\dfrac{1}{\sqrt{3}}$ and θ is in Q II.

   Therefore, $\theta = 150°$.

   $z_2 = 2(\cos 150° + i \sin 150°)$ in trigonometric form

   $z_1 z_2 = 2 \cdot 2[\cos(60° + 150°) + i \sin(60° + 150°)]$
   $\qquad = 4(\cos 210° + i \sin 210°)$

   $4(\cos 210° + i \sin 210°) = 4(-\dfrac{\sqrt{3}}{2} - i \cdot \dfrac{1}{2})$
   $\qquad\qquad\qquad\qquad = -2\sqrt{3} - 2i$

13. $z_1 z_2 = (1 + i)(4i)$
$$= 4i + 4i^2$$
$$= 4i + 4(-1)$$
$$= -4 + 4i$$

$z_1 = 1 + i$ where $x = 1$, $y = 1$, and $r = \sqrt{1^2 + 1^2} = \sqrt{2}$.
Also, $\tan \theta = 1$ and θ is in Q I.
Therefore, $\theta = 45°$.

In trigonometric form, $z_1 = \sqrt{2}(\cos 45° + i \sin 45°)$.

$z_2 = 4i$ where $x = 0$, $y = 4$, and $r = \sqrt{0^2 + 4^2} = 4$.
Also, $\theta = 90°$.

In trigonometric form, $z_2 = 4(\cos 90° + i \sin 90°)$.

$z_1 z_2 = \sqrt{2} \cdot 4[\cos(45° + 90°) + i \sin(45° + 90°)]$
$$= 4\sqrt{2}(\cos 135° + i \sin 135°)$$

$4\sqrt{2}(\cos 135° + i \sin 135°) = 4\sqrt{2}(-\frac{\sqrt{2}}{2} + i \cdot \frac{\sqrt{2}}{2})$
$$= -4 + 4i$$

17. $[2(\cos 10° + i \sin 10°)]^6 = 2^6[\cos(6 \cdot 10°) + i \sin(6 \cdot 10°)]$
$$= 64(\cos 60° + i \sin 60°)$$
$$= 64(\frac{1}{2} + i \cdot \frac{\sqrt{3}}{2})$$
$$= 32 + 32i\sqrt{3}$$

21. $[3(\cos 60° + i \sin 60°)]^4 = 3^4[\cos(4 \cdot 60°) + i \sin(4 \cdot 60°)]$
$$= 81(\cos 240° + i \sin 240°)$$
$$= 81[-\frac{1}{2} + i(-\frac{\sqrt{3}}{2})]$$
$$= -\frac{81}{2} - \frac{81\sqrt{3}}{2}i$$

25. First we write $1 + i$ in trigonometric form:

$z = 1 + i$ where $x = 1$, $y = 1$, and $r = \sqrt{1^2 + 1^2} = \sqrt{2}$.

Also, $\tan \theta = 1$ and θ is in Q I.
Therefore, $\theta = 45°$.

In trigonometric form, $z = \sqrt{2}(\cos 45° + i \sin 45°)$.

$z^4 = (\sqrt{2})^4[\cos(4 \cdot 45°) + i \sin(4 \cdot 45°)]$
$$= 4(\cos 180° + i \sin 180°)$$
$$= 4[-1 + i(0)]$$
$$= -4$$

29. First we write 1 - i in trigonometric form:

$z = 1 - i$ where $x = 1$, $y = -1$, and $r = \sqrt{1^2 + (-1)^2} = \sqrt{2}$.

Also, $\tan \theta = -1$ and θ is in Q IV.
Therefore, $\theta = 315°$.

In trigonometric form, $z = \sqrt{2}(\cos 315° + i \sin 315°)$.

$$z^6 = (\sqrt{2})^6 [\cos(6 \cdot 315°) + i \sin(6 \cdot 315°)]$$
$$= 8(\cos 1890° + i \sin 1890°)$$
$$= 8(\cos 90° + i \sin 90°)$$
$$= 8(0 + i \cdot 1)$$
$$= 8i$$

33. $\dfrac{20(\cos 75° + i \sin 75°)}{5(\cos 40° + i \sin 40°)} = \dfrac{20}{5} [\cos(75° - 40°) + i \sin(75° - 40°)]$

$$= 4(\cos 35° + i \sin 35°)$$

37. $\dfrac{4(\cos 90° + i \sin 90°)}{8(\cos 30° + i \sin 30°)} = \dfrac{4}{8} [\cos(90° - 30°) + i \sin(90° - 30°)]$

$$= \frac{1}{2} (\cos 60° + i \sin 60°)$$

41. $\dfrac{z_1}{z_2} = \dfrac{\sqrt{3} + i}{2i} \cdot \dfrac{i}{i}$

$$= \frac{i\sqrt{3} + i^2}{2i^2}$$

$$= \frac{-1 + i\sqrt{3}}{-2}$$

$$= \frac{1}{2} - \frac{\sqrt{3}}{2} i$$

$z_1 = \sqrt{3} + i$ where $x = \sqrt{3}$, $y = 1$, and $r = \sqrt{(\sqrt{3})^2 + 1^2} = 2$.

Also, $\tan \theta = \dfrac{1}{\sqrt{3}}$ and θ is in Q I.

Therefore, $\theta = 30°$.

In trigonometric form, $z_1 = 2(\cos 30° + i \sin 30°)$.

$z_2 = 2i$ where $x = 0$, $y = 2$, and $r = \sqrt{0^2 + 2^2} = 2$.
Also, $\theta = 90°$.

In trigonometric form, $z_2 = 2(\cos 90° + i \sin 90°)$.

$\dfrac{z_1}{z_2} = \dfrac{2(\cos 30° + i \sin 30°)}{2(\cos 90° + i \sin 90°)}$

$$= \frac{2}{2} [\cos(30° - 90°) + i \sin(30° - 90°)]$$

$$= \cos(-60°) + i \sin(-60°)$$

$$\cos(-60°) + i \sin(-60°) = \frac{1}{2} + i(-\frac{\sqrt{3}}{2})$$

$$= \frac{1}{2} - \frac{\sqrt{3}}{2} i$$

45. $\frac{z_1}{z_2} = \frac{8}{-4} = -2$

$z_1 = 8$ where $x = 8$, $y = 0$, and $r = 8$. Also, $\theta = 0°$.

In trigonometric form, $z_1 = 8(\cos 0° + i \sin 0°)$.

$z_2 = -4$ where $x = -4$, $y = 0$, and $r = 4$. Also, $\theta = 180°$.

In trigonometric form, $z_2 = 4(\cos 180° + i \sin 180°)$.

$\frac{z_1}{z_2} = \frac{8(\cos 0° + i \sin 0°)}{4(\cos 180° + i \sin 180°)}$

$= \frac{8}{4} [\cos(0° - 180°) + i \sin(0° - 180°)]$

$= 2[\cos(-180°) + i \sin(-180°)]$

$2[\cos(-180°) + i \sin(-180°)] = 2[-1 + i(0)]$
$= -2$

49. Let $z_1 = 1 + i\sqrt{3}$, where $x = 1$, $y = \sqrt{3}$, and $r = \sqrt{1^2 + (\sqrt{3})^2} = 2$.
Also, $\tan \theta = \sqrt{3}$ and θ is in Q I.
Therefore, $\theta = 60°$.

In trigonometric form, $z_1 = 2(\cos 60° + i \sin 60°)$.

Let $z_2 = \sqrt{3} - i$, where $x = \sqrt{3}$, $y = -1$, and $r = \sqrt{(\sqrt{3})^2 + (-1)^2} = 2$.

Also, $\tan \theta = -\frac{1}{\sqrt{3}}$ and θ is in Q IV.

Therefore, $\theta = 330°$.

In trigonometric form, $z_2 = 2(\cos 330° + i \sin 330°)$.

Let $z_3 = 1 - i\sqrt{3}$, where $x = 1$, $y = -\sqrt{3}$, and $r = \sqrt{1^2 + (-\sqrt{3})^2} = 2$.
Also, $\tan \theta = -\sqrt{3}$ and θ is in Q IV.
Therefore, $\theta = 300°$.

In trigonometric form, $z_3 = 2(\cos 300° + i \sin 300°)$.

Now, we find $\dfrac{(z_1)^4(z_2)^2}{(z_3)^3}$:

$$\frac{(z_1)^4(z_2)^2}{(z_3)^3} = \frac{2^4[\cos(4 \cdot 60°) + i\sin(4 \cdot 60°)]2^2[\cos(2 \cdot 330°) + i\sin(2 \cdot 330°)]}{2^3[\cos(3 \cdot 300°) + i\sin(3 \cdot 300°)]}$$

$$= \frac{16 \cdot 4[\cos(240° + 660°) + i\sin(240° + 660°)]}{8(\cos 900° + i\sin 900°)}$$

$$= \frac{64}{8} \cdot [\cos(900° - 900°) + i\sin(900° - 900°)]$$

$$= 8(\cos 0° + i\sin 0°)$$

$$= 8(1 + i \cdot 0)$$

$$= 8$$

53. $w^4 = [2(\cos 15° + i\sin 15°)]^4$

$$= 2^4[\cos(4 \cdot 15°) + i\sin(4 \cdot 15°)]$$

$$= 16(\cos 60° + i\sin 60°)$$

$$= 16(\tfrac{1}{2} + i \cdot \tfrac{\sqrt{3}}{2})$$

$$= 8 + 8i\sqrt{3}$$

57. First, we write $\sqrt{3} - i$ in trigonometric form:

$z = \sqrt{3} - i$ where $x = \sqrt{3}$, $y = -1$, and $r = \sqrt{(\sqrt{3})^2 + (-1)^2} = 2$.

Also, $\tan \theta = -\dfrac{1}{\sqrt{3}}$ and θ is in Q IV.

Therefore, $\theta = 330°$.

In trigonometric form, $z = 2(\cos 330° + i\sin 330°)$.

$z^{-1} = 2^{-1}[\cos(-1)(330°) + i\sin(-1)(330°)]$

$$= \tfrac{1}{2}[\cos(-330°) + i\sin(-330°)]$$

$$= \tfrac{1}{2}[\tfrac{\sqrt{3}}{2} + i \cdot \tfrac{1}{2}]$$

$$= \frac{\sqrt{3}}{4} + \frac{1}{4}i$$

61. $\sin \dfrac{A}{2} = \sqrt{\dfrac{1 - \cos A}{2}}$

$$= \sqrt{\frac{1 - (-\tfrac{1}{3})}{2}}$$

$$= \sqrt{\frac{2}{3}}$$

$$= \frac{\sqrt{6}}{3}$$

65. If $\cos A = -\dfrac{1}{3}$ and A is in Q II, then

$$\sin A = \sqrt{1 - \cos^2 A}$$

$$= \sqrt{1 - \dfrac{1}{9}} = \sqrt{\dfrac{8}{9}} = \dfrac{2\sqrt{2}}{3}$$

Also, $\tan A = \dfrac{\sin A}{\cos A} = \dfrac{\dfrac{2\sqrt{2}}{3}}{\dfrac{1}{3}} = 2\sqrt{2}$

Therefore, $\tan 2A = \dfrac{2\tan A}{1 - \tan^2 A}$

$$= \dfrac{2(2\sqrt{2})}{1 - (2\sqrt{2})^2}$$

$$= \dfrac{4\sqrt{2}}{1 - 8}$$

$$= -\dfrac{4\sqrt{2}}{7}$$

69. $|\vec{g}| = \sqrt{(170)^2 + (28)^2 - 2(170)(28)\ \cos 68°}$

$$= \sqrt{28,900 + 784 - 9520(0.3746)}$$

$$= \sqrt{26,118}$$

$$= 160 \text{ mph (to 2 significant digits)}$$

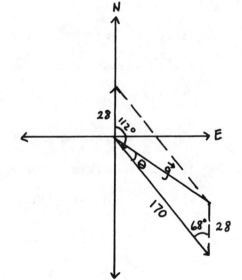

$\dfrac{\sin \theta}{28} = \dfrac{\sin 68°}{160}$

$\sin \theta = \dfrac{28 \sin 68°}{160}$

$\sin \theta = \dfrac{28(0.9272)}{160}$

$\sin \theta = 0.1623$

$\theta = 9°$ (to the nearest degree)

The true course is $112° - \theta = 112° - 9° = 103°$.
The ground speed is 160 miles per hour.

Problem Set 8.4

1. The 2 square roots will be

$$w_k = 4^{\frac{1}{2}}[\cos \frac{30° + 360°k}{2} + i \sin \frac{30° + 360°k}{2}] \text{ for } k = 0, 1$$

$$= 2[\cos(15° + 180°k) + i \sin(15° + 180°k)]$$

Replacing k with 0 and 1, we have

when k = 0, $w_0 = 2(\cos 15° + i \sin 15°)$
when k = 1, $w_1 = 2(\cos 195° + i \sin 195°)$

5. The 2 square roots will be

$$w_k = 49^{\frac{1}{2}}[\cos \frac{180° + 360°k}{2} + i \sin \frac{180° + 360°k}{2}] \text{ for } k = 0, 1$$

$$= 7[\cos(90° + 180°k) + i \sin(90° + 180°k)]$$

Replacing k with 0 and 1, we have

when k = 0, $w_0 = 7(\cos 90° + i \sin 90°)$
when k = 1, $w_1 = 7(\cos 270° + i \sin 270°)$

9. In trigonometric form, $4i = 4(\cos 90° + i \sin 90°)$.

The 2 square roots of 4i will be

$$w_k = 4^{\frac{1}{2}}[\cos \frac{90° + 360°k}{2} + i \sin \frac{90° + 360°k}{2}] \text{ for } k = 0, 1$$

$$= 2[\cos(45° + 180°k) + i \sin(45° + 180°k)]$$

Replacing k with 0 and 1, we have

when k = 0, $w_0 = 2(\cos 45° + i \sin 45°)$

$$= 2(\frac{\sqrt{2}}{2} + i \cdot \frac{\sqrt{2}}{2})$$

$$= \sqrt{2} + i\sqrt{2}$$

when k = 1, $w_1 = 2(\cos 225° + i \sin 225°)$

$$= 2[-\frac{\sqrt{2}}{2} + i \cdot (-\frac{\sqrt{2}}{2})]$$

$$= -\sqrt{2} - i\sqrt{2}$$

13. In trigonometric form, $1 + i\sqrt{3} = 2(\cos 60° + i \sin 60°)$

The 2 square roots of $1 + i\sqrt{3}$ will be

$$w_k = 2^{\frac{1}{2}}[\cos \frac{60° + 360°k}{2} + i \sin \frac{60° + 360°k}{2}] \text{ for } k = 0, 1$$

$$= \sqrt{2}[\cos(30° + 180°k) + i \sin(30° + 180°k)]$$

Replacing k with 0 and 1, we have

when k = 0, $w_0 = \sqrt{2}(\cos 30° + i \sin 30°)$

$$= \sqrt{2}(\frac{\sqrt{3}}{2} + i \cdot \frac{1}{2})$$

$$= \frac{\sqrt{6}}{2} + \frac{i\sqrt{2}}{2}$$

when k = 1, $w_1 = \sqrt{2}(\cos 210° + i \sin 210°)$

$$= \sqrt{2}[-\frac{\sqrt{3}}{2} + i \cdot (-\frac{1}{2})]$$

$$= -\frac{\sqrt{6}}{2} - \frac{i\sqrt{2}}{2}$$

17. First, we write $4\sqrt{3} + 4i$ in trigonometric form:

$x = 4\sqrt{3}$, $y = 4$, and $r = \sqrt{(4\sqrt{3})^2 + 4^2} = 8$.

Also, $\tan \theta = \frac{4}{4\sqrt{3}} = \frac{1}{\sqrt{3}}$ and θ is in Q I.

Therefore, $\theta = 30°$.

$4\sqrt{3} + 4i = 8(\cos 30° + i \sin 30°)$ in trigonometric form.

The 3 cube roots will be

$$w_k = 8^{1/3}[\cos \frac{30° + 360°k}{3} + i \sin \frac{30° + 360°k}{3}] \text{ for } k = 0, 1, 2$$

$$= 2[\cos(10° + 120°k) + i \sin(10° + 120°k)]$$

Replacing k with 0, 1, and 2, we have

when k = 0, $w_0 = 2(\cos 10° + i \sin 10°)$
when k = 1, $w_1 = 2(\cos 130° + i \sin 130°)$
when k = 2, $w_2 = 2(\cos 250° + i \sin 250°)$

21. In trigonometric form, $64i = 64(\cos 90° + i \sin 90°)$

The 3 cube roots will be

$$w_k = 64^{1/3}[\cos \frac{90° + 360°k}{3} + i \sin \frac{90° + 360°k}{3}] \text{ for } k = 0, 1, 2$$

$$= 4[\cos(30° + 120°k) + i \sin(30° + 120°k)]$$

Replacing k with 0, 1, and 2, we have

when k = 0, $w_0 = 4(\cos 30° + i \sin 30°)$
when k = 1, $w_1 = 4(\cos 150° + i \sin 150°)$
when k = 2, $w_2 = 4(\cos 270° + i \sin 270°)$

25. $x^4 - 16 = 0$
    $x^4 = 16$ The solutions to this equation are the 4 fourth roots of 16.

In trigonometric form, $16 = 16(\cos 0° + i \sin 0°)$

The 4 fourth roots will be

$$w_k = 16^{\frac{1}{4}}[\cos \frac{0° + 360°k}{4} + i \sin \frac{0° + 360°k}{4}] \text{ for } k = 0, 1, 2, 3$$

$$w_k = 2[\cos 90°k + i \sin 90°k]$$

Replacing k with 0, 1, 2, and 3, we have

when $k = 0$, $w_0 = 2(\cos 0° + i \sin 0°)$
            $= 2(1 + i \cdot 0) = 2$

when $k = 1$, $w_1 = 2(\cos 90° + i \sin 90°)$
            $= 2(0 + i \cdot 1) = 2i$

when $k = 2$, $w_2 = 2(\cos 180° + i \sin 180°)$
            $= 2(-1 + i \cdot 0) = -2$

when $k = 3$, $w_3 = 2(\cos 270° + i \sin 270°)$
            $= 2[0 + i(-1)] = -2i$

29. The 5 fifth roots will be

$$w_k = (10^5)^{1/5}[\cos \frac{15° + 360°k}{5} + i \sin \frac{15° + 360°k}{5}] \text{ for } k = 0, 1, 2, 3, 4$$

$$= 10[\cos(3° + 72°k) + i \sin(3° + 72°k)]$$

Replacing k with 0, 1, 2, 3, and 4, we have

when $k = 0$, $w_0 = 10(\cos 3° + i \sin 3°)$
            $= 10[0.999 + i(0.052)]$
            $= 9.99 + 0.52i$

when $k = 1$, $w_1 = 10(\cos 75° + i \sin 75°)$
            $= 10[0.259 + i(0.966)]$
            $= 2.59 + 9.66i$

when $k = 2$, $w_2 = 10(\cos 147° + i \sin 147°)$
            $= 10[-0.839 + i(0.545)]$
            $= -8.39 + 5.45i$

when $k = 3$, $w_3 = 10(\cos 219° + i \sin 219°)$
            $= 10[-0.777 + i(-0.629)]$
            $= -7.77 - 6.29i$

when $k = 4$, $w_4 = 10(\cos 291° + i \sin 291°)$
            $= 10[0.358 + i(-0.934)]$
            $= 3.58 - 9.34i$

33. Applying the quadratic formula, we have

$$x^2 = \frac{2 \pm \sqrt{(-2)^2 - 4(1)(4)}}{2(1)}$$

$$= \frac{2 \pm \sqrt{-12}}{2} = \frac{2 \pm 2i\sqrt{3}}{2} = 1 \pm i\sqrt{3}$$

That is, $x^2 = 1 + i\sqrt{3}$ or $x^2 = 1 - i\sqrt{3}$.

First, we find the 2 square roots of $1 + i\sqrt{3}$:

$1 + i\sqrt{3} = 2(\cos 60° + i \sin 60°)$ in trigonometric form

The 2 square roots will be

$$w_k = 2^{\frac{1}{2}}[\cos \frac{60° + 360°k}{2} + i \sin \frac{60° + 360°k}{2}] \text{ for } k = 0, 1$$

$$= \sqrt{2}[\cos(30° + 180°k) + i \sin(30° + 180°k)]$$

Replacing k with 0 and 1, we have

when $k = 0$, $w_0 = \sqrt{2}(\cos 30° + i \sin 30°)$
when $k = 1$, $w_1 = \sqrt{2}(\cos 210° + i \sin 210°)$

Second, we find the 2 square roots of $1 - i\sqrt{3}$:

$1 - i\sqrt{3} = 2(\cos 300° + i \sin 300°)$ in trigonometric form

The 2 square roots will be

$$w_k = 2^{\frac{1}{2}}[\cos \frac{300° + 360°k}{2} + i \sin \frac{300° + 360°k}{2}] \text{ for } k = 0, 1$$

$$= \sqrt{2}[\cos(150° + 180°k) + i \sin(150° + 180°k)]$$

Replacing k with 0 and 1, we have

when $k = 0$, $w_0 = \sqrt{2}(\cos 150° + i \sin 150°)$
when $k = 1$, $w_1 = \sqrt{2}(\cos 330° + i \sin 330°)$

37. $y = -2 \sin(-3x) = 2 \sin 3x$ because sine is an odd function.
The graph is a sine curve with amplitude = 2 and period = $\frac{2\pi}{3}$.

41. The graph is a sine curve with

Amplitude = 3

Period = $\frac{2\pi}{\pi/3} = 6$

Phase shift = $\frac{\pi/3}{\pi/3} = 1$

45. $\quad s = \dfrac{1}{2}(a + b + c)$ Formula for half-perimeter

$\quad\quad = \dfrac{1}{2}(2.3 + 3.4 + 4.5)$ Substitute known values

$\quad\quad = 5.1$ Simplify

$S = \sqrt{s(s - a)(s - b)(s - c)}$ Formula for Area of triangle

$\quad = \sqrt{5.1(5.1 - 2.3)(5.1 - 3.4)(5.1 - 4.5)}$ Substitute known values

$\quad = \sqrt{5.1(2.8)(1.7)(0.6)}$ Simplify

$\quad = \sqrt{14.5656}$

$\quad = 3.8 \text{ feet}^2$ Round to 2 significant digits

Problem Set 8.5

13. All points of the form (2, 60° + 360°k), where k = an integer, will name the point (2, 60°). For example,

     if k = -1, we have (2, -300°).

Also, all points of the form (-2, 240° + 360°k), where k = an integer, will name the point (2, 60°). For example,

     if k = 0, we have (-2, 240°).
     if k = -1, we have (-2, -120°).

17. All points of the form (-3, 30° + 360°k), where k = an integer, will name the point (-3, 30°). For example,

     if k = -1, we have (-3, -330°).

Also, all points of the form (3, 210° + 360°k), where k = an integer, will name the point (-3, 30°). For example,

     if k = 0, we have (3, 210°).
     if k = -1, we have (3, -150°).

21. $\quad x = r \cos \theta \quad\quad$ and $\quad y = r \sin \theta$
$\quad x = 3 \cos 270° \quad\quad\quad y = 3 \sin 270°$
$\quad x = 3(0) \quad\quad\quad\quad\quad y = 3(-1)$
$\quad x = 0 \quad\quad\quad$ and $\quad y = -3$

$\quad\quad (3, 270°) = (0, -3)$

25. $\quad x = r \cos \theta \quad\quad$ and $\quad y = r \sin \theta$
$\quad x = -4\sqrt{3} \cos 30° \quad\quad\quad y = -4\sqrt{3} \sin 30°$

$\quad x = -4\sqrt{3}\left(\dfrac{\sqrt{3}}{2}\right) \quad\quad\quad y = -4\sqrt{3}\left(\dfrac{1}{2}\right)$

$\quad x = -6 \quad\quad\quad\quad\quad y = -2\sqrt{3}$

$\quad\quad (-4\sqrt{3}, 30°) = (-6, -2\sqrt{3})$

29. $r = \sqrt{x^2 + y^2}$ and $\tan \theta = \dfrac{y}{x}$

     $= \sqrt{(-2\sqrt{3})^2 + (2)^2}$ $\tan \theta = \dfrac{2}{-2\sqrt{3}}$

     $= \sqrt{12 + 4}$ $\tan \theta = -\dfrac{1}{\sqrt{3}}$

     $= 4$ $\theta = 150°$ or $330°$

Since $(-2\sqrt{3},\ 2)$ is in Q II, one solution is $(4,\ 150°)$.

33. $r = \sqrt{x^2 + y^2}$ and $\tan \theta = \dfrac{y}{x}$

     $= \sqrt{(-\sqrt{3})^2 + (-1)^2}$ $\tan \theta = \dfrac{-1}{-\sqrt{3}}$

     $= \sqrt{3 + 1}$ $\tan \theta = \dfrac{1}{\sqrt{3}}$

     $= 2$ $\theta = 30°$ or $210°$

Since $(-\sqrt{3},\ -1)$ is in Q III, one solution is $(2,\ 210°)$.

37. $r = \sqrt{x^2 + y^2}$ and $\tan \theta = \dfrac{y}{x}$

     $= \sqrt{(-1)^2 + (2)^2}$ $\tan \theta = \dfrac{2}{-1}$

     $= \sqrt{1 + 4}$ $\theta = 116.6°$ or $296.6°$

     $= \sqrt{5}$

Since $(-1,\ 2)$ is in Q II, one solution is $(\sqrt{5},\ 116.6°)$.

41. $r^2 = 9$

   $x^2 + y^2 = 9$ Substitute $x^2 + y^2$ for r^2

45. $r^2 = 4 \sin 2\theta$

     $r^2 = 4(2 \sin \theta \cos \theta)$ Double-angle identity

     $r^2 = 8\left(\dfrac{y}{r}\right)\left(\dfrac{x}{r}\right)$ Substitute $\dfrac{y}{r}$ for $\sin \theta$ and $\dfrac{x}{r}$ for $\cos \theta$

     $r^2 = \dfrac{8xy}{r^2}$ Multiply

     $r^4 = 8xy$ Multiply both sides by r^2

  $(x^2 + y^2)^2 = 8xy$ Substitute $x^2 + y^2$ for r^2

49. $\qquad x - y = 5$
   $r \cos \theta - r \sin \theta = 5$ Substitute $r \cos \theta$ for x and $r \sin \theta$ for y
   $\quad r(\cos \theta - \sin \theta) = 5$ Factor out r

53. $x^2 + y^2 = 6x$
   $\qquad r^2 = 6r \cos \theta$ Substitute r^2 for $x^2 + y^2$ and $r \cos \theta$ for x
   $\qquad r = 6 \cos \theta$ Divide both sides by r

57. The graph is a sine curve with amplitude = 6.

61. Let $y_1 = 4$ (a horizontal line) and $y_2 = 2 \sin x$ (a sine curve with amplitude 2). Graph y_1, y_2, and $y = y_1 + y_2$ on the same coordinate axes.

Problem Set 8.6

1.

θ	$r = 6 \cos \theta$	(r, θ)
$0°$	$r = 6 \cos 0° = 6$	$(6, 0°)$
$45°$	$r = 6 \cos 45° = 4.2$	$(4.2, 45°)$
$90°$	$r = 6 \cos 90° = 0$	$(0, 90°)$
$135°$	$r = 6 \cos 135° = -4.2$	$(-4.2, 135°)$
$180°$	$r = 6 \cos 180° = -6$	$(-6, 180°)$
$225°$	$r = 6 \cos 225° = -4.2$	$(-4.2, 225°)$
$270°$	$r = 6 \cos 270° = 0$	$(0, 270°)$
$315°$	$r = 6 \cos 315° = 4.2$	$(4.2, 315°)$

5. $\qquad r = 3$
   $\pm\sqrt{x^2 + y^2} = 3$ Substitute $\pm\sqrt{x^2 + y^2}$ for r
   $\quad x^2 + y^2 = 9$ Square both sides

The graph is a circle with center at the origin or pole and radius of 3.

9. First, we sketch $y = 3 \sin x$:

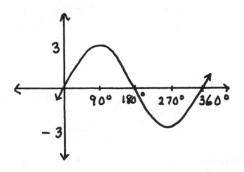

Next, we note the relationship between variations in x (or θ) and the corresponding variations in y (or r).

154

Variations in x (or θ)	Corresponding Variations in y (or r)
0° to 90°	0 to 3
90° to 180°	3 to 0
180° to 270°	0 to -3
270° to 360°	-3 to 0

Then, we sketch the graph using this relationship.

13. First, we sketch y = 2 + 4 cos x:

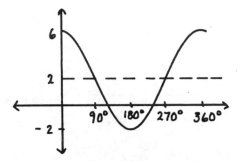

Next, we note the relationship between variations in x (or θ) and the corresponding variations in y (or r).

Variations in x (or θ)	Corresponding Variations in y (or r)
0° to 90°	6 to 2
90° to 180°	2 to -2
180° to 270°	-2 to 2
270° to 360°	2 to 6

Then, we sketch the graph using this relationship.

17. $r^2 = 9 \sin 2\theta$ is a lemniscate or two-leaved rose. (See Figure 12 in text-book.) The endpoints of the leaves are $(\sqrt{a}, 45°) = (3, 45°)$ and $(-\sqrt{a}, 45°) = (-3, 45°)$.

21. $r = 4 \cos 3\theta$ is a three-leaved rose. (See Figure 11 in textbook). The end-points of the leaves are $(a, 0°) = (4, 0°)$, $(a, 120°) = (4, 120°)$, and $(a, 240°) = (4, 240°)$.

25. $x^2 + y^2 = 6x$
      $r^2 = 6r \cos \theta$ Substitute r^2 for $x^2 + y^2$ and $r \cos \theta$ for x
      $r = 6 \cos \theta$ Divide both sides by r

The graph is a circle with center at (3, 0°) and radius = 3.

29. $r(2 \cos \theta + 3 \sin \theta) = 6$
    $2r \cos \theta + 3r \sin \theta = 6$ Multiply
          $2x + 3y = 6$ Substitute x for $r \cos \theta$ and y for $r \sin \theta$

The graph is a line through (0, 2) and (3, 0).

33.
$$r = 4 \sin \theta$$
$$r^2 = 4r \sin \theta \qquad \text{Multiply both sides by } r$$
$$x^2 + y^2 = 4y \qquad \text{Substitute } x^2 + y^2 \text{ for } r^2 \text{ and } y \text{ for } r \sin \theta$$
$$x^2 + y^2 - 4y = 0 \qquad \text{Subtract } 4y \text{ from both sides}$$
$$x^2 + y^2 - 4y + 4 = 4 \qquad \text{Complete the square by adding 4 to both sides}$$
$$x^2 + (y - 2)^2 = 4 \qquad \text{Factor}$$

The graph is a circle with center at (0, 2) and radius = 2.

37. Let $y_1 = \sin x$ (the basic sine curve) and $y_2 = -\cos x$ (a cosine curve reflected about the x-axis). Graph y_1, y_2, and $y = y_1 + y_2$ on the same corrdinate axes.

41. Let $y_1 = 3 \sin x$ (a sine curve with amplitude = 3) and $y_2 = \cos 2x$ (a cosine curve with period = $\frac{2\pi}{2} = \pi$). Graph y_1, y_2, and $y = y_1 + y_2$ on the same coordinate axes.

Chapter 8 Test

1. $\sqrt{-25} = \sqrt{25}\ i = 5i$

5. $(6 - 3i) + [(4 - 2i) - (3 + i)] = 6 - 3i + 4 - 2i - 3 - i$
$$= (6 + 4 - 3) + (-3i - 2i - i)$$
$$= 7 - 6i$$

9. $(8 + 5i)(8 - 5i) = 64 + 40i - 40i - 25i^2$
$$= 64 - 25(-1)$$
$$= 64 + 25$$
$$= 89$$

13. (a) $|3 + 4i| = \sqrt{(3)^2 + (4)^2}$
$$= \sqrt{25}$$
$$= 5$$

   (b) $-(3 + 4i) = -3 - 4i$

   (c) The conjugate of $3 + 4i$ is $3 - 4i$.

17. $8(\cos 330° + i \sin 330°) = 8[\frac{\sqrt{3}}{2} + i(-\frac{1}{2})]$
$$= 4\sqrt{3} - 4i$$

21. $5i$ lies on the positive y-axis.
Therefore, $r = 5$ and $\theta = 90°$.
$5i = 5(\cos 90° + i \sin 90°)$

25. $[2(\cos 10° + i \sin 10°)]^5 = 2^5[\cos(5 \cdot 10°) + i \sin(5 \cdot 10°)]$
$$= 32(\cos 50° + i \sin 50°)$$

29. Using the quadratic formula, we have
$$x^2 = \frac{2\sqrt{3} \pm \sqrt{12 - 4(1)(4)}}{2(1)}$$
$$= \frac{2\sqrt{3} \pm \sqrt{-4}}{2}$$
$$= \frac{2\sqrt{3} \pm 2i}{2}$$
$$= \sqrt{3} \pm i$$

Therefore, $x^2 = \sqrt{3} + i$ or $x^2 = \sqrt{3} - i$.

First, we will find the 2 square roots of $\sqrt{3} + i$:

$\sqrt{3} + i = 2(\cos 30° + i \sin 30°)$ in trigonometric form

The 2 square roots will be

$$w_k = 2^{\frac{1}{2}}[\cos \frac{30° + 360°k}{2} + i \sin \frac{30° + 360°k}{2}] \text{ for } k = 0, 1$$

$$= \sqrt{2}[\cos(15° + 180°k) + i \sin(15° + 180°k)]$$

For $k = 0$, $w_0 = \sqrt{2}(\cos 15° + i \sin 15°)$
For $k = 1$, $w_1 = \sqrt{2}(\cos 195° + i \sin 195°)$

Next, we will find the 2 square roots of $\sqrt{3} - i$:

$\sqrt{3} - i = 2(\cos 330° + i \sin 330°)$ in trigonometric form

The 2 square roots will be

$$w_k = 2^{\frac{1}{2}}[\cos \frac{330° + 360°k}{2} + i \sin \frac{330° + 360°k}{2}] \text{ for } k = 0, 1$$

$$= \sqrt{2}[\cos(165° + 180°k) + i \sin(165° + 180°k)]$$

For $k = 0$, $w_0 = \sqrt{2}(\cos 165° + i \sin 165°)$
For $k = 1$, $w_1 = \sqrt{2}(\cos 345° + i \sin 345°)$

33. $r = \sqrt{x^2 + y^2}$ and $\tan \theta = \frac{y}{x}$

$\quad = \sqrt{(-3)^2 + (3)^2}$ $\qquad \tan \theta = \frac{3}{-3} = -1$

$\quad = \sqrt{18}$ $\qquad\qquad\qquad \theta = 135° \text{ or } 315°$

$\quad = 3\sqrt{2}$

Since $(-3, 3)$ is in Q II, one solution is $(3\sqrt{2}, 135°)$

37. $\qquad\qquad\quad x + y = 2$
$\quad r \cos \theta + r \sin \theta = 2 \qquad$ Substitute $r \cos \theta$ for x and $r \sin \theta$ for y
$\quad r(\cos \theta + \sin \theta) = 2 \qquad$ Factor out r

41. First, we sketch $y = 4 + 2 \cos x$:

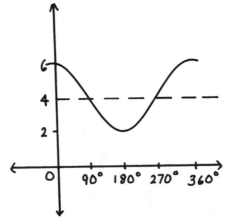

Next, we note the relationship between variations in x (or θ), and the corresponding variations in y (or r).

Variations in x (or θ)	Corresponding Variations in y (or r)
0° to 90°	6 to 4
90° to 180°	4 to 2
180° to 270°	2 to 4
270° to 360°	4 to 6

Then, we sketch the graph using this relationship.

APPENDIX LOGARITHMS

Problem Set A.1

17. $\log_3 x = 2$
       $x = 3^2$ Change to exponential form
       $x = 9$ Simplify

21. $\log_x 4 = 2$

       $x^2 = 4$ Change to exponential form
       $x = \pm 2$ Take square root of both sides
       $x = 2$ x must be positive

25. $y = \log_{1/3} x$

       $x = \left(\dfrac{1}{3}\right)^y$ Change to exponential form

x	9	3	1	$\frac{1}{3}$	$\frac{1}{9}$
y	-2	-1	0	1	2

Graph these ordered pairs and join them with a smooth curve.

29. $y = \log_{25} 125$

      $25^y = 125$ Change to exponential form

      $(5^2)^y = 5^3$ Write 25 and 125 as powers of 5

       $2y = 3$ If $b^m = b^n$, then $m = n$

       $y = \dfrac{3}{2}$ Divide both sides by 2

33. $\log_3 3 = 1$ (See example 7 in textbook.)

37. $\log_3(\log_6 6) = \log_3 1$ $\log_b b = 1$

                 $= 0$ $\log_b 1 = 0$

41. $pH = -\log_{10}[H^+]$
         $= -\log_{10}(10^{-7})$ Substitute given values
         $= -(-7)$ $\log_b b^x = x$

         $= 7$ Simplify

45. $M = \log_{10} T$ Formula given in textbook

      $= \log_{10} 100$ Substitute given values

      $= \log_{10} 10^2$ Write 100 as a power of 10

      $= 2$ $\log_b b^x = x$

Problem Set A.2

1. $\log_3 4x = \log_3 4 + \log_3 x$ Property 1

5. $\log_2 y^5 = 5 \log_2 y$ Property 3

9. $\log_6 x^2 y^3 = \log_6 x^2 + \log_6 y^3$ Property 1

      $= 2 \log_6 x + 3 \log_6 y$ Property 3

13. $\log_b \dfrac{xy}{z} = \log_b (xy) - \log_b z$ Property 2

      $= \log_b x + \log_b y - \log_b z$ Property 1

17. $\log_{10} \dfrac{x^2 y}{\sqrt{z}} = \log_{10} (x^2 y) - \log_{10} z^{\frac{1}{2}}$ Property 2

      $= \log_{10} x^2 + \log_{10} y - \log_{10} z^{\frac{1}{2}}$ Property 1

      $= 2 \log_{10} x + \log_{10} y - \dfrac{1}{2} \log_{10} z$ Property 3

21. $\log_b \sqrt[3]{\dfrac{x^2 y}{z^4}} = \log_b \left(\dfrac{x^2 y}{z^4}\right)^{1/3}$ Write as exponent

      $= \log_b \dfrac{x^{2/3} y^{1/3}}{z^{4/3}}$ Property of exponents

      $= \log_b (x^{2/3} y^{1/3}) - \log_b z^{4/3}$ Property 2

      $= \log_b x^{2/3} + \log_b y^{1/3} - \log_b z^{4/3}$ Property 1

      $= \dfrac{2}{3} \log_b x + \dfrac{1}{3} \log_b y - \dfrac{4}{3} \log_b z$ Property 3

25. $2 \log_3 x - 3 \log_3 y = \log_3 x^2 - \log_3 y^3$ Property 3

      $= \log_3 \dfrac{x^2}{y^3}$ Property 2

29. $3 \log_2 x + \frac{1}{2} \log_2 y - \log_2 z = \log_2 x^3 + \log_2 y^{\frac{1}{2}} - \log_2 z$ Property 3

$$= \log_2 x^3 y^{\frac{1}{2}} - \log_2 z \qquad \text{Property 1}$$

$$= \log_2 \frac{x^3 y^{\frac{1}{2}}}{z} \text{ or } \log_2 \frac{x^3 \sqrt{y}}{z} \qquad \text{Property 2}$$

33. $\frac{3}{2} \log_{10} x - \frac{3}{4} \log_{10} y - \frac{4}{5} \log_{10} z$

$$= \log_{10} x^{3/2} - \log_{10} y^{3/4} - \log_{10} z^{4/5} \qquad \text{Property 3}$$

$$= \log_{10} x^{3/2} - (\log_{10} y^{3/4} + \log_{10} z^{4/5}) \qquad \text{Introduce parentheses}$$

$$= \log_{10} x^{3/2} - \log_{10} (y^{3/4} z^{4/5}) \qquad \text{Property 1}$$

$$= \log_{10} \left(\frac{x^{3/2}}{y^{3/4} z^{4/5}} \right) \qquad \text{Property 2}$$

37. $\log_3 x - \log_3 2 = 2$

$\log_3 \frac{x}{2} = 2$ Property 2

$\frac{x}{2} = 3^2$ Change to exponential form

$\frac{x}{2} = 9$ Simplify

$x = 18$ Multiply both sides by 2

41. $\log_3 (x + 3) - \log_3 (x - 1) = 1$

$\log_3 \left(\frac{x + 3}{x - 1} \right) = 1$ Property 2

$\frac{x + 3}{x - 1} = 3^1$ Change to exponential form

$x + 3 = 3x - 3$ Multiply both sides by $x - 1$

$-2x = -6$ Solve the linear equation

$x = 3$

45. $\log_8 x + \log_8 (x - 3) = \frac{2}{3}$

$\log_8 x(x - 3) = \frac{2}{3}$ Property 1

$x(x - 3) = 8^{2/3}$ Change to exponential form

$x(x - 3) = 4$ Simplify

$x^2 - 3x - 4 = 0$ Put in standard form

$(x - 4)(x + 1) = 0$ Factor left side

$x - 4 = 0 \text{ or } x + 1 = 0$ Set each factor = 0

$x = 4 \qquad x = -1$ Solve and check

Solution is 4   -1 does not check

49. $M = 0.21(\log_{10} a - \log_{10} b)$

    $= 0.21(\log_{10} 1 - \log_{10} 10^{-12})$ Substitute given values

    $= 0.21[0 - (-12)]$ $\log_b 1 = 0$ and $\log_b b^x = x$

    $= 0.21(12)$ Simplify

    $= 2.52$ Multiply

  $M = 0.21 \log_{10} \left(\dfrac{a}{b}\right)$

    $= 0.21 \log_{10} \left(\dfrac{1}{10^{-12}}\right)$ Substitute given values

    $= 0.21 \log_{10} 10^{12}$ Simplify

    $= 0.21(12)$ $\log_b b^x = x$

    $= 2.52$ Multiply

53. $\log_{10} A = \log_{10} 100(1.06)^t$

    $= \log_{10} 100 + \log_{10}(1.06)^t$ Property 1

    $= \log_{10}(10^2) + t \log_{10}(1.06)$ Property 3

    $= 2 + t \log_{10}(1.06)$ $\log_b b^x = x$

Problem Set A.3

1. Calculator: 378 $\boxed{\log}$
   Answer: 2.5775

5. Calculator: 3780 $\boxed{\log}$
   Answer: 3.5775

9. Calculator: 37,800 $\boxed{\log}$
   Answer: 4.5775

13. Calculator: 2010 $\boxed{\log}$
    Answer: 3.3032

17. Calculator: 2.8802 $\boxed{10^x}$ (or $\boxed{\text{inv}}$ $\boxed{\log}$)
    Answer: 759

21. Calculator: 3.1553 $\boxed{10^x}$
    Answer: 1430

25. $pH = -\log[H^+]$
    $= -\log(6.5 \times 10^{-4})$
    $= -\log(0.00065)$
    $= -(-3.2)$
    $= 3.2$

29. $M = \log T$

    $5.5 = \log T$ Substitute given value

    $T = 316{,}000$ Calculator: 5.5 $\boxed{10^x}$

33. $M = \log T$

    $6.5 = \log T$ Substitute given value

    $T = 3{,}160{,}000$ Calculator: 6.5 $\boxed{10^x}$

Now, we compare this to our answer from #29 and find it is 10 times as large.

37. $\log(1 - r) = \dfrac{1}{t} \log \dfrac{w}{p}$

    $\log(1 - r) = \dfrac{1}{5} \log \dfrac{5750}{7550}$ Substitute given values

    $\log(1 - r) = \dfrac{1}{5} (-0.1183)$ Use $\boxed{\log}$ on calculator

    $\log(1 - r) = -0.0237$ Multiply

        $1 - r = 0.947$ Use $\boxed{10^x}$ on calculator

           $r = 0.053$ Solve linear equation

           $r = 5.3\%$ Change to a percent

41. $\ln e^5 = 5 \ln e$ Property 3

        $= 5(1)$ $\log_e e = 1$

        $= 5$ Multiply

45. $\ln 10e^{3t} = \ln 10 + \ln e^{3t}$ Property 1

           $= \ln 10 + 3t \ln e$ Property 3

           $= \ln 10 + 3t(1)$ $\ln e = 1$

           $= \ln 10 + 3t$ Multiply

49. $\ln 15 = \ln(5 \cdot 3)$

          $= \ln 5 + \ln 3$ Property 1

          $= 1.6094 + 1.0986$ Substitute given values

          $= 2.7080$ Add

53. $\ln 9 = \ln 3^2$

        $= 2 \ln 3$ Property 3

        $= 2(1.0986)$ Substitute given value

        $= 2.1972$ Multiply

164

1. $3^x = 5$

      $\log 3^x = \log 5$ Take log of both sides

      $x \log 3 = \log 5$ Property 3

      $x = \dfrac{\log 5}{\log 3}$ Divide both sides by log 3

      $x = \dfrac{0.6990}{0.4771}$ Use $\boxed{\log}$ on calculator

      $x = 1.4650$ Divide

5. $5^{-x} = 12$

      $\log 5^{-x} = \log 12$ Take log of both sides

      $-x \log 5 = \log 12$ Property 3

      $x = -\dfrac{\log 12}{\log 5}$ Divide both sides by $-\log 5$

      $x = -\dfrac{1.0792}{0.6990}$ Use $\boxed{\log}$ on calculator

      $= -1.5440$ Divide

9. $3^{2x+1} = 2$

      $\log 3^{2x+1} = \log 2$ Take log of both sides

      $(2x + 1)\log 3 = \log 2$ Property 3

      $2x + 1 = \dfrac{\log 2}{\log 3}$ Divide both sides by log 3

      $2x = \dfrac{\log 2}{\log 3} - 1$ Subtract 1 from both sides

      $x = \dfrac{1}{2}\left[\dfrac{\log 2}{\log 3} - 1\right]$ Multiply both sides by $\dfrac{1}{2}$

      $x = \dfrac{1}{2}\left[\dfrac{0.3010}{0.4771} - 1\right]$ Use $\boxed{\log}$ on calculator

      $x = \dfrac{1}{2}(0.6309 - 1)$ Divide

      $x = \dfrac{1}{2}(-0.3691)$ Subtract

      $x = -0.1845$ Multiply

13. $A = P(1 + \frac{r}{n})^{nt}$ where $P = \$5000$, $r = .12$, $n = 1$, $t = 10$

     $= 5000(1 + .12)^{10}$ Substitute given values

     $= 5000(1.12)^{10}$ Add

     $= 5000(3.1058)$ Calculator: 1.12 $\boxed{y^x}$ 10 $\boxed{=}$

     $= \$15,529$ Multiply

17.

     $A = P(1 + \frac{r}{n})^{nt}$ where $P = \$500$, $A = \$1000$, $r = .06$, $n = 2$

     $1000 = 500(1 + \frac{.06}{2})^{2t}$ Substitute given values

     $1000 = 500(1.03)^{2t}$ Simplify

     $(1.03)^{2t} = 2$ Divide both sides by 500

     $\log(1.03)^{2t} = \log 2$ Take log of both sides

     $2t \log 1.03 = \log 2$ Property 3

     $t = \frac{\log 2}{2 \log 1.03}$ Divide both sides by 2 log 1.03

     $t = \frac{0.3010}{0.0257}$ Use $\boxed{\log}$ on calculator

     $t = 11.7$ years   Divide

21. $\log_8 16 = \frac{\log 16}{\log 8}$ 25. $\log_7 15 = \frac{\log 15}{\log 7}$

     $= \frac{1.2041}{0.9031}$ $= \frac{1.1761}{0.8451}$

     $= 1.3333$ $= 1.3917$

29. Calculator: 345 $\boxed{\ln}$ 33. Calculator: 10 $\boxed{\ln}$

     Answer: 5.8435   Answer: 2.3026

37. $32,000 \, e^{0.05t} = 64,000$ Substitute given value

     $e^{0.05t} = 2$ Divide both sides by 32,000

     $0.05t = \ln 2$ Change to logarithmic form

     $t = \frac{\ln 2}{0.05}$ Divide both sides by .05

     $t = \frac{0.6931}{0.05}$ Use $\boxed{\ln}$ on calculator

     $t = 13.9$ years or toward the end of 2001

41. $A = Pe^{rt}$

$\dfrac{A}{P} = e^{rt}$ Divide both sides by P

$rt = \ln \dfrac{A}{P}$ Change to logarithmic form

$t = \dfrac{1}{r} \ln \dfrac{A}{P}$ Multiply both sides by $\dfrac{1}{r}$

45. $A = P(1 - r)^{t}$

$\dfrac{A}{P} = (1 - r)^{t}$ Divide both sides by P

$\log(1 - r)^{t} = \log \dfrac{A}{P}$ Take log of both sides

$t \log(1 - r) = \log A - \log P$ Property 3 and Property 2

$t = \dfrac{\log A - \log P}{\log(1 - r)}$ Divide both sides by $\log(1 - r)$